Maria Liziane Souza Silva

The Battle of the Rubber:

Maria Liziane Souza Silva

The Battle of the Rubber:

Northeastern migrants - memory and imagination

ScienciaScripts

Imprint

Any brand names and product names mentioned in this book are subject to trademark, brand or patent protection and are trademarks or registered trademarks of their respective holders. The use of brand names, product names, common names, trade names, product descriptions etc. even without a particular marking in this work is in no way to be construed to mean that such names may be regarded as unrestricted in respect of trademark and brand protection legislation and could thus be used by anyone.

Cover image: www.ingimage.com

This book is a translation from the original published under ISBN 978-613-9-73545-7.

Publisher:
Sciencia Scripts
is a trademark of
Dodo Books Indian Ocean Ltd. and OmniScriptum S.R.L publishing group

120 High Road, East Finchley, London, N2 9ED, United Kingdom
Str. Armeneasca 28/1, office 1, Chisinau MD-2012, Republic of Moldova, Europe
Printed at: see last page
ISBN: 978-620-7-87225-1

To ALL the rubber tappers, children and grandchildren of Northeastern descent living in the state of Acre, especially my mother, Maria de Lourdes Souza Silva; my grandmother Guiomar Medeiros Marques; uncles and aunts who are the fruit of this process, for fighting and believing that the days would be better.

ACKNOWLEDGEMENTS

> *To my God, for without him nothing would be possible, and to the many people who helped me directly and indirectly to realise this project:*

> *To my husband Paulo César Barros Pereira, also a Master's student at the PPGG, for always encouraging me and making my dream his own. Thank you for your support, dedication and for being my foundation. I love you!*

> *To my children Isabela S. Barros Pereira and Paulo Arthur S. Barros Pereira for the love they have given me. You have been my strength, my vigour, my hope for better days. It was for you and to you. I love you deeply!*

> *To my dear mum, Maria de Lourdes Souza Silva, for taking the time to tell me about our northeastern background;*

> *To my siblings who have supported and encouraged me, especially during this first year of my Master's programme. I love you and thank you so much for your strength and dedication!*

> *To my father, Suterlande Marques, for his promptness in taking me to the various contacts interviewed in Mâncio Lima;*

> *To my dear little voice, Guiomar Medeiro Marques, who, at the age of 90, readily gave me an account of her career;*

> *To Mr Murilo de Lima, a northeastern rubber soldier who lives in Mâncio Lima and who, at the age of 98, gave me an extraordinary account of his life. Congratulations on your lucidity and vigour!*

> *To the various interviewees, children, grandchildren and wives of Northeasterners: Mrs Maria Helena and her husband Manoel Cláudio; Mrs Maria dos Anjos; Mr Adalton Monteiro de Lima; My dear Aunt Dê - who left us five months after giving me the interview; Madrinha Rosa; Dona Osmarina; Uncle Lú Mourão; All these residents of Mâncio Lima Acre. And also Mr Divaldo Alves de Souza, 91, and his wife Oscarina Alves de Souza, 89, both currently living in Rio Branco, Acre.*

> *To my beloved undergraduate professors, Prof. Dr. DF. Maria de Jesus Morais; Prof. DF. Jose Alves; Prof. DF DF Karina Furini; ProF. DF Lucilene da Silva Almeida, all teachers on the Geography programme at the Federal University of Acre, for their support, encouragement and help over the years. You are my mirror!*

> *To the Federal University of Rondônia Foundation, especially to my friends from the 2016/2 Master's programme in Geography that I have built up during this short time, and to all the professors of the PPGG course. Thank you to the "cafofo do Acre", a group of people from Acre studying for their Masters in this programme;*

> *The Rondônia Foundation for the Support of the Development of Scientific and Technological Actions and Research in the State of Rondônia (FAPERO) for the grant that enabled me to dedicate myself fully to the activities of the PPGG/UNIR. This benefit was essential to my career*

and staying on the programme, since I left my hometown of Rio Branco Acre for the city of Porto Velho/RO without financial support. I would therefore like to express my gratitude to this institution, which helped me in my academic training and was fundamental to the realisation of this research.

> *To the Group for Studies and Research in Geography, Women and Social Gender Relations - GEPGÊNERO. Thank you for sharing your experiences and knowledge during trips, events and/or fieldwork. Thank you for believing in me!*

> *To dear Prof Dr Josué da Costa Silva, for his great hospitality and kindness in welcoming us to his home;*

> *And without a doubt, dear ProF. DF. Maria das Graças Silva Nascimento Silva, who has helped me throughout this time and from whom I have learnt so much.*

To all of you, thank you very much!

SUMMARY

The Amazon/Acre region, once home to an abundance of rubber trees, was an attraction for thousands of people who came mainly from the north-east of Brazil at the height of the rubber boom. These movements included those of the late 19th and mid-20th centuries, the period of the Second World War. In the federal government's propaganda, this region was portrayed as a "land of plenty and victory", an attraction for the recruitment of thousands of men as soldiers, the so-called "rubber soldiers". The aim of this work is to reflect, through the stories told by Northeasterners living in the state of Acre and/or their wives and children, who are also the fruits of this process since they experienced the difficulties of the rubber plantations on a daily basis, on the reasons that led them to migrate; how the call was made; the promises made by the government; how they were transported from Ceara to the Amazonian rubber plantations; the dangers and adversities faced by the new soldier within the rubber plantations and whether the outcome of the whole experience was satisfactory in the "new life" project. Discussing these migrants highlights the 140 years of migration from the northeast to the Amazon, which began in 1877, enriching the theoretical discourse on the phenomenon. In order to understand the experiences of these migrants, we used the phenomenological approach, as it is a method that takes into account perceptions, subjectivities, emotions and their affectivities. During the interview, it allowed for an exchange of empathetic feelings, so as to be able to put oneself in the other person's shoes, in the capacity of being able to feel the other person's pain and understand them as if they were one. It was by focusing on these sensations, on valuing these sensitivities, listening to each particularity, that phenomenology was chosen. The main technique adopted in this study was the oral source, which over the centuries has proved to be the greatest human source for preserving and disseminating knowledge. The narratives of these subjects reveal that the new region had no connection with the reality conveyed by the adverts; on the contrary, their speeches deconstruct every fallacy implanted by the posters of the time. The region was very different and the days were more than difficult. Nobody got rich, nobody's life was transformed by the wealth of rubber. Everything was utopia. The dream of wealth and a return to a better life was smoke, an illusion. However, they absorbed the new space as their own. They became so intimate with it that even though they reveal all the limitations they experienced, even today they still opt for the place "of the forest", of the rubber plantations, of nostalgia. In truth, these migratory movements were key points in the social, economic and cultural formation of what would become "Acre". They were essential to the emergence of a new place, forming a new identity, shown by the work, way of life, experiences, sufferings and hopes of these migrants who arrived here in search of work and "easy money".

Key words: Rubber tapping; Northeasterners; Amazonia/Acre.

SUMMARY

PRESENTATION

I started my academic life late, almost at the age of thirty. But inside me, the flame to acquire scientific knowledge burned brightly and never went out. I'm from the interior of Acre, from the far west of the state, the municipality of Mâncio Lima, for those of you who are familiar: Japiim. The land where "the wind makes the curve"; the westernmost land in the country; the land of the "Serra do Môa"; the land of manioc flour; the land of the rubber tapper Mâncio Lima; the land bathed by the Moa River with its countless tributaries, cut by igarapés and igapós.

I come from the land of countless rubber tappers, descendants of people from Ceará who arrived in the late 19th and mid-20th centuries. A land that welcomed these thousands, even when it was not yet a Brazilian territory. They were the heroic and anonymous pioneers who, at the sacrifice of their lives, explored and conquered the Amazon, linking it to future generations with all its immense potential for natural wealth.

It was there that I heard countless stories, tales, from my ancestors, rubber tappers from Ceará, of the struggle, of coming to this region, of the suffering, the anguish, the loneliness, the immense difficulties, the dangers, but ....also of the parties, the partying, the typical foods, customs, prayers, sayings, language, nostalgia. I thought it was all very beautiful, without, however, reflecting on it.

Like them, I also went through the migration process, internally of course, to the capital Rio Branco, back in the 1990s. I took with me the same dreams as my ancestors: to have a better life, to seek new possibilities, new opportunities. I knew it would be difficult to make a fortune, and I certainly didn't. I worked from an early age, like my parents, grandparents and uncles, because I saw work as a prospect of financial change. It certainly didn't come, I wasn't blessed with it, but that's not the point. Because deep down I felt what was fundamental: study! I was always passionate about the discoveries it would give me, the new worlds I would discover, the invisible becoming visible.

It was with this love of study that I entered university, albeit late. It was a celebration, I was the first child of six siblings to go to university, something unusual in our family. It wasn't my favourite course, but I fell in love with Geography. I turned this door into the great opportunity of my life. However, as a bush bug, rusty from my studies, I felt the weight of the new journey. I didn't give up, I embraced it. It was at this point that the "scales began to fall off". Because that's what studying does: it takes the blinders off. Entering university brought me good results and more than expected results. The knowledge and experience I've acquired has been immense. Participating as a volunteer and scholarship holder in the Tutorial Education Programme - PET Geography was fundamental to my maturing.

That was the moment when I chose to do my monographic work on what permeated my imagination as a child: "migration from Ceará to the Amazon". More than an assignment, it was going back in time, reliving all my imagination and discovering much more than I had ever known. I was also fortunate to have the help and partnership of professors from the Federal University of Acre - UFAC, especially Prof.ª Dr.ª Maria de Jesus Morais, who guided me and taught me to see further.

That's why when I decided to take on another stage of study, the biggest challenge I'd achieved so far, the Master's in Geography, was just a test. In 2016, entering the Masters and Doctorate Postgraduate Programme at the Federal University of Rondônia Foundation - PPGG/UNIR, was my greatest achievement. No one in my family, neither first nor second degree, had ever studied for a master's degree. A master's degree was unimaginable in the family.

In this context, given all of the above: the history of my experience, the knowledge I brought with me and my trajectory, the choice of the Master's research project could not have been any different when it came to discussing, now in more depth, how this migration took place and what the results of this process were. My own family was my source of research. There were also those friends and acquaintances, those recommended by the family, who helped me a lot. There are so many of them that I didn't have time or conditions to access them all. But everyone, when they heard about it, was very happy to know that someone cared about their story, their trajectory. So the aim here is to valorise these values, the stories, the knowledge, the lived experiences, the relationships with nature, which I sometimes notice are being lost over time.

I arrived at UNIR's Master's programme in Geography in August 2016. We started the programme with a beautiful inaugural lecture and a welcome for the new students. The following week I took my first course, "Epistemology of Geography", taught by Professor Dr Josué Costa. The course was concentrated and divided into three modules. It was very intense. At that moment I felt the first impact, I realised that my vision was limited and that it was now necessary to see beyond the superficial. I was taken by surprise, shaken, and it required me to take a more refined look at geographical science. The various authors, their concepts and the exchanges of information debated in class with the teacher and other classmates brought me a whirlwind of information I had never had before. I learnt a lot and was able to enrich my theoretical base.

Other subjects such as: Amazonian Populations and Sustainability; Research Methods and Techniques in Geography and Geoprocessing and Digital Cartography also added other perspectives and knowledge that made me get drunk on geographical science. All these subjects gave me a new angle and a specific, refined look at each one, corroborating the answers to my research problem.

I also have to emphasise the enormous learning I had from the different fields carried out during my two-year Master's degree, which gave me new meanings of the Amazonian space. Learning from the GEPgênero research group, led by my supervisor, Professor Dr Maria das Graças Silva, affectionately known as "Professor Gracinha", was undoubtedly extremely important for my theoretical growth.

All the experiences I've had, the activities I've done, the new friendships I've made during this period, the mishaps that have occurred, the financial difficulties, have all been steps I've had to take to reach this final moment. The Master's period was a watershed for me, it served as a breakthrough in the construction of new geographical thinking. I was rebuilt! However, I continue in this incessant quest for learning, because I know that it never stops.

The idea behind this work is that it can be a starting point for other scientific reflections and that it can contribute to other academic research.

INTRODUCTION

The occupation of the Brazilian Amazon by non-Indians is marked by different migratory movements. These movements include those at the end of the 19th century and in the middle of the 20th century, in the 1940s, when the rubber boom was in full swing. They were made up of waves of people who came mainly from the north-east of Brazil, who wandered into the lush green forests of the Amazon, a region rich in biodiversity and abundant in "white gold", as the latex extracted from the rubber tree is called.

At first, at the end of the 19th century, the international conditions arising from the Second Industrial Revolution that was taking place in the USA, when the automobile industry became the centre of gravity of the system and, consequently, demanded the search for raw materials for the manufacture of tyres, caused the rubber market economy to boom. The Amazon, a region abundant in this raw material, became attractive to thousands who came to cut rubber. It is estimated that from 1850 to 1900 the population of the Amazon valley increased tenfold. The figures reach 158,125 northeasterners alone, around 20 per cent of the Amazonian population at the time. In Acre, men from the backlands of Ceará, Pernambuco, Paraíba and Rio Grande do Norte also began arriving from 1877 onwards (MARTINELLO, 2004).

The peak of this economy was in 1912, when it reached the highest number of tonnes of rubber exported abroad. But after this period, the Amazon collapsed when an Asian competitor appeared, taking the rubber plantations to a catastrophic level, with Amazonian rubber reaching minimum export values, a loss of almost 92 per cent (MARTINELLO, 2004). This decline lasted for long and bitter years and the rubber tappers, who had arrived in search of a new life through the profit they would make from their labour, fell into oblivion, abandoned by those who had recruited them. The great Eldorado, as this region was seen, had become a disappointment.

But from the 1940s onwards, this economy was once again rekindled during the "Battle of the Rubber", during the Second World War, as a result of the commitments made by Brazil to the US through the "Washington Agreement". Now we have a new scenario, with new prospects for rekindling an economy that had once enjoyed splendour. It was against this backdrop that another migration to the Amazon took place, mostly by northeasterners. Once again, thousands saw this as an opportunity to make a fortune, given the intense advertising that daily filled the newspapers, radio stations and posters scattered around the cities of the northeast showing the Amazon as the place "of victory"; of "fortune", as the perfect place for those who wanted to make money. In addition to the high expectations of this area, as presented by the press at the time, there was also the fact that those called up for the war could choose between serving as soldiers on the war front or serving in the rubber plantations as "rubber soldiers". Many of them chose to serve in the rubber plantations instead of the war, because they realised that here they would have a better chance of survival.

There is no collection, archive or site provided by the state with the names of these workers. When you look for documentation, there is a significant lack of numbers. It is therefore not possible to know exactly how many people were sent into the forest and how many died in the process. In the book *Amazônia - Um Pouco Antes e Além Depois* (1977), the Amazonian Samuel Benchimol counts the arrival of 50,500 men between 1943 and 1944, at the height of conscription, and in their luggage, around 19,760 women, of whom almost nothing is known.

It was the beginning of the "Battle of the Rubber", a name given to them because it was a time of war and, although it wasn't an armed dispute, they were labelled soldiers because they were serving their country by cutting rubber. The state of Acre was one of the Amazonian states that received these people.

But in the midst of such a wealth of fauna and flora, these thousands were faced with an inhospitable land, full of dangers and diseases typical of the region itself. This would be the scene of huge disputes between peoples, witness to the greed of the wealthiest, the hard work of those who believed in it so much, the loneliness, the dreams, the plans of thousands of people who arrived here in search of a better life, a place to work and change the reality of their financial situation, a region for the northeasterners.

The people who emigrated were those who had left their parents, fleeing poverty, in search of the great Eldorado, often getting lost forever in the confines of the Amazon jungle, never to hear from their loved ones again. The conditions that welcomed them were as devastating as possible. Benchimol (1977) reports that it was common for supplies to be unavailable for transport upriver when navigation was possible, and for foodstuffs to rot in Belém and Manaus, having arrived there when the upper rivers were dry, resulting in a year of deprivation without any production. The men were supplied as "things", without knowing if the rubber plantations were prepared to receive them or how they were going to feed them. It is estimated that more than 30,000 died in the Amazon by the end of the war, in addition to the hundreds who couldn't resist the long journey by lorry and boat.

For those who resisted, few knew exactly what they would face in the forest and many ended up victims of disease and attacks by wild animals such as snakes and jaguars, but none of them were prepared for the action of a species of animal known for its greed: man himself! The emperors of this region, owners of rubber plantations. Those who gambled that they would have assistance and better living conditions were faced with a situation similar to the conditions found at the end of the 19th century, that of semi-slavery.

With the end of the Second World War in 1945, England resumed its production of latex from Asia and the Brazilian product was no longer of interest to the international market. With no demand, the rubber plantations began to be abandoned once again, as well as their workers, who were left to fend for themselves.

It is in this set of details that the importance of the chosen theme *"THE BATTLE OF THE RUBBER: Northeastern migrants - memory and imaginary" comes into play - a* starting point for understanding northeastern migration to the Amazon. An Amazon formed in the image of Ceará and the rubber plantation, key to the social, economic and cultural formation of the Amazon/Acre area. Furthermore, the discussion of these migrants highlights the 140 years of northeastern migration to the Amazon, which began in 1877, enriching the theoretical discourse on the phenomenon.

Given this context, some questions arise: how was this large labour force summoned? How did these rubber tappers act in the face of adversity, whether due to the danger of attack by animals and/or Indians, or diseases within the rubber plantations? What was the role/reality of women rubber tappers at that time? Did anyone benefit from the wealth promised by the government through rubber tapping? Did they really achieve the so-called "new life" that the posters propagated? What is your assessment of this process?

The aim is to understand, through the accounts of these Northeasterners living in the state of Acre and/or their wives and children, who are also the fruits of this process, their daily lives, experiences, difficulties and limitations within the rubber plantations. The idea is to be able to look at it from another angle and show that there may be other stories than those told in books.

Study Area: The State of Acre

The spatial scope of this study is limited to the state of Acre, located in the south of the Amazon, a region that is still home to thousands of these rubber tappers. In order to locate these individuals within the state, three regions were selected that were the entry points for these migrants, limited to the three main rivers that cut through the state: the Acre River in the far east of the state where the capital Rio Branco is located, the Tarauacá River in the centre of the state on the banks of the city of Tarauacá and in the far west the Juruá River near the municipality of Mâncio Lima. These rivers are tributaries of the Amazon River, where rubber tappers travelled from the ports of Belém to Acre during the golden age of rubber. These three municipalities were chosen because they cover more of the state.

After 1877, when explorers arrived, recruited by rubber tappers, to work extracting latex in this region due to the high prices of rubber on the international market and, consequently, the interference of monopoly capital, which imposed a new logic on mercantile accumulation, they found a region known as "Aquiri", a name given by the Apurinãs Indians, the first inhabitants of the region, which means "river of alligators". The origin of the name Acre was given by the descendants of matutos from Ceará who transcribed the name from the indigenous dialect, giving rise to the name Acre. The territory, formerly belonging to Bolivia and Peru, was occupied by Brazilians over the decades (CALIXTO, 2003; BRASIL, 2018).

Acre, one of Brazil's 27 states, is located in the western portion of the country's northern region, in the western Amazon, between longitudes 66°38' WGr and 74°00' WGr and latitudes 7°07' S and 11°08' S. It shares international borders with Peru and Bolivia and state borders with Rondônia and the state of Amazonas (ACRE, 2008). The area of the territorial unit is 164,123.737 km^2 . It is the 16th largest state in Brazil. The last IBGE census (2010) revealed a population of 733,559 inhabitants, with a population density of 4.47 inhabitants/km^2 , but for 2018 the population estimate is 869,265 people (BRASIL, 2018). The state currently has 22 municipalities and its capital is the city of Rio Branco.

Historically, Acre's economy has been based on plant extraction, especially the exploitation of rubber, which was responsible for the non-indigenous settlement of the region. Wood is currently the state's main export product, and it is also a major producer of Brazil nuts, açaí and copaiba oil. The main agricultural activities include manioc, maize, rice and sugar cane (ACRE, 2008).

Acre has two economic hubs: the Juruá river valley, which has the city of Cruzeiro do Sul as its main urban centre; and the Acre river valley, which as well as being home to the capital Rio Branco, is the most industrialised, has the highest degree of mechanisation and modernisation in the countryside, and the greatest potential in agricultural activities (ACRE, 2008). But despite this, the state contributed just 0.2 per cent to Brazil's Gross Domestic Product (GDP), and even though it grew by 4.4 per cent in 2014, it still ranks among the country's lowest GDPs, alongside Roraima and Amapá. The state's GDP is made up of services (68.1%), industry (14.7%) and agriculture (17.2%) (VIDAL and ALVES, 2017).

Located in the south of the Brazilian Amazon, Acre has a unique history. In its 16 million hectares of rainforest, it has the greatest biodiversity on earth and many of its inhabitants still live in the forest. Among them are Indians, from the 32 indigenous reserves in the state, 14 different nations who, despite the many cultural losses following contact with non-indigenous people, still preserve their traditions, cultures and rituals. There are also communities in the state that are organised on the basis of a family production unit that uses rivers as its main means of transport and the forest itself as a source of food (ACRE, 2017).

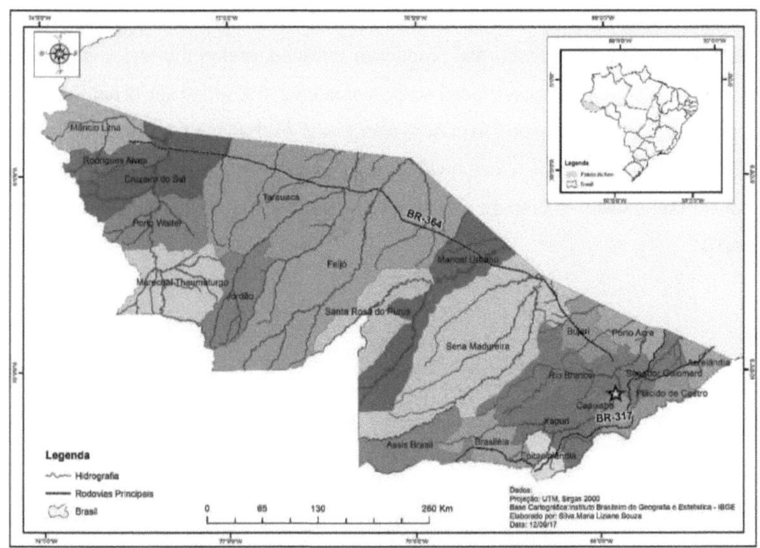

Map 01: Location of the state of Acre and its municipalities
Org: Maria Liziane S. Silva, 2017.

Acre is considered a new state. On 15 June 1962, through Law 4.070, it was elevated from the status of Federal Territory to the category of State. At that time, the municipalities of Cruzeiro do Sul, Tarauacá, Feijó, Sena Madureira, Xapuri, Brasileia and Rio Branco already existed. With the promulgation of the State Constitution of Acre on 1º March 1963, several new municipalities were to be created, but the lack of territorial delimitation and financial resources meant that they were not actually established until 1976, when they began to enjoy autonomous municipal administration. These municipalities were: Mâncio Lima; Assis Brasil; Manuel Urbano, Plácido de Castro and Senador Guiomard. In 1992, a further 10 municipalities were created: Acrelândia; Bujarí; Capixaba; Epitaciolândia; Jordão; Marechal Thaumaturgo; Porto Walter; Porto Acre; Rodrigues Alves and Santa Rosa do Purus, thus making up 22 municipalities (ACRE, 2008).

However, as a form of administrative gestational improvement, today the state is divided into two mesoregions: Vale do Acre and Vale do Juruá, and also into 5 development regions: Alto Acre, Baixo Acre, Purus, Tarauacá/Envira and Juruá, which follow the distribution of the hydrographic basins of the main rivers in Acre. Currently, 71 per cent of the population is concentrated in urban areas, with the capital, Rio Branco, accounting for 58 per cent of this urban population (Acre, 2009).

The westernmost of Brazil's states, Acre still preserves a large part of its forest area, around 90 per cent. Until the 1970s, more than half of Acre's population lived in a strong relationship with the forest, as rubber tappers, river dwellers or indigenous people, inspiring the name known today as "forest peoples" or "traditional Amazonian peoples". (ACRE, 2010)

The life of this Amazonian man has always been full of challenges, especially when the economic base was latex extraction. When this production declined, during the first and second heyday of rubber, the Amazon region always faced serious problems of a strong social nature, facts that are narrated in this study by those who live in Acre and/or who experienced this situation.In this sense, having outlined our objectives, the structuring of this research had to go through some stages. We will now show the paths taken to prepare and complete this research, organised according to the theoretical map 01:

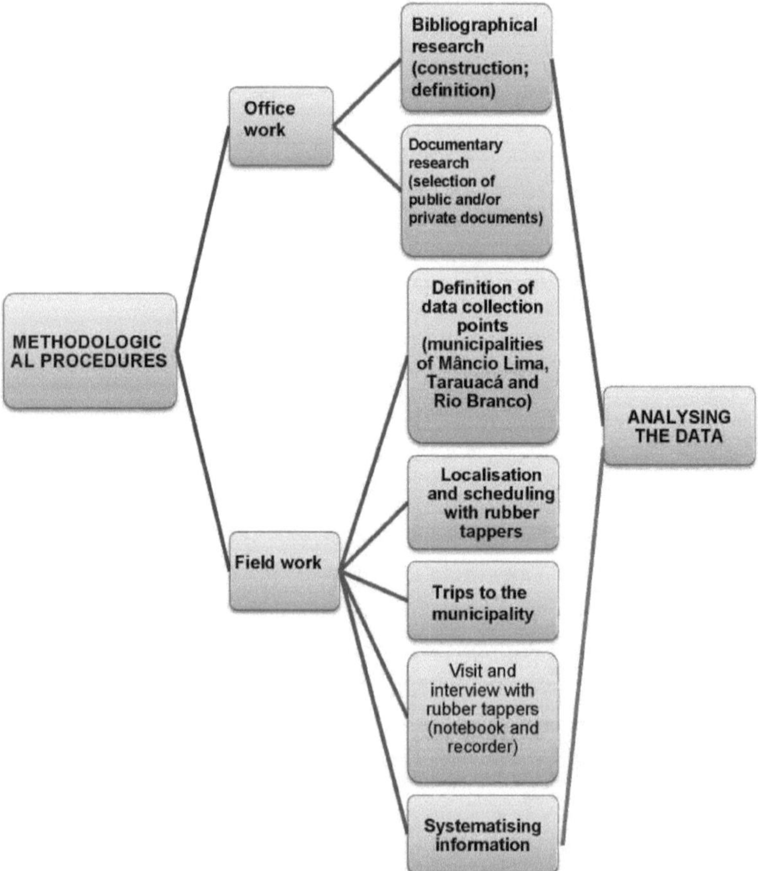

Figure 01: Methodological overview
Office work, 2018.
Prepared by: SILVA, Maria Liziane Souza.

The description of the main methodological stages for understanding this geographical phenomenon is summarised below:

J <u>Bibliographical and documentary research:</u> theoretical and conceptual review of the theme of northeastern migration in the Amazon (space, migration, history of Acre, First and Second

Rubber Battle in the Amazon), based on articles, books, dissertations, theses and documents from public and private institutions made available via the internet or by the institution itself, such as: The Rondônia Postgraduate Master's and Doctoral Programme in Geography PPGG; the Federal University of Acre - UFAC; the Acre Technology Foundation -FUNTAC; the Brazilian Institute of Statistics - IBGE and the Acre State Planning Secretariat - SEPLAN; <u>Fieldwork: the</u> aim was to analyse the narratives of some Northeasterners, their wives or children who still live in the state of Acre and who lived through the migration process during the Second World War and/or, being their descendants, worked cutting rubber with their parents and spouses. Through these narratives we sought to give a voice to these subjects, who have always been made invisible, and to be able to know other truths beyond what is put down in official books, believing that there are other stories, other plots that we can discover about the whole migration process and the experiences within the rubber plantations. The field research took place between February and May 2017;

J <u>Definition of data collection points</u>: The definition of the collection points was defined seeking to have a greater territorial reach in the state. In this way, the idea was to collect data on the three main access routes that these rubber tappers used during the rubber battle, which were the rivers: Acre (in the city of Rio Branco); Tarauacá (in the city of Tarauacá) and Juruá (in Mâncio Lima). After locating them in these municipalities, the interviews were scheduled and then the trips to the municipalities were made. However, it also happened that some interviews were carried out without prior scheduling due to referrals from neighbours and relatives who led us to other rubber tappers.

Semi-structured interviews were carried out with the rubber tappers' wives and children, as these narratives express spontaneity and shed light on the mentalities of their daily lives and experiences (QUEIROZ, 1991). To this end, a sequential script of open and/or closed questions was used. As for the research subjects, 14 people were interviewed, including men and women aged between 70 and 93, living in the municipalities of Mâncio Lima, Tarauacá and Rio Branco, with the aim of recovering the story of their trajectory through memory. The interviews often began informally, when they acted with great hospitality and kindness at the reception.

As well as using a field notebook to make the necessary notes, a mobile phone voice recorder was also used to record the interviews, and also to record images of the interviewees. The use of the recorder technique indicates that "the transcription carried out by the researcher himself can therefore enrich the document and its information" (QUEIROZ, 1991, p.87).

J <u>Systematisation of information:</u> By systematising the data obtained in the field, the public documents analysed, the institutions and bodies investigated, and correlating them with the conceptual theoretical references, we sought to analyse and understand the migratory phenomenon of northeasterners that manifested itself in the Amazon during the rubber economy.

In this context, the proposal to dissert on this theme, together with the intended epistemological approach, allows us to understand how this phenomenon has interfered, modified and added values to the new Amazonian living space, which can be explained through the lens of phenomenological geography.Thus, this dissertation is structured in four chapters:

In the first chapter: *"GEOGRAPHICAL SPACE AND MIGRATION: Theoretical assumptions and understanding in geographical analysis".* This is a theoretical and methodological approach by authors who discuss the issues at hand: geographical space and the term migration. The aim, above all, was to provide some conceptualisations that contribute to the concept of geographical space, and then to discuss the concepts that guide the term migration, highlighting the "whys" and the reasons why they happened.

The second chapter: "THE *PHENOMENOLOGICAL METHOD: Getting back to things".* This presents the methodological part, detailing the order and step-by-step process for carrying out the research, a way of obtaining a result based on a theory. The phenomenological method was chosen in this study because it helps to understand the subjectivities of the narratives of the selected subjects. The techniques used included bibliographical research, oral sources, snowballing and the Discourse of the Collective Subject (DSC). These methodologies have their own ideologies and epistemological positions regarding *the* object studied.

Third chapter: *"THE GREAT ELDORADO: From the two great internal migrations to the constitution of the state of Acre".* This chapter presents the two great northeastern migrations that took place in the Amazon during the height of the rubber boom, which contributed to the formation of the territory of Acre. The aim was to understand the internal and external context in which Brazil/Amazon was inserted, resulting in a population shift never seen before.

The fourth chapter: *"THE VOICE OF THE "INVISIBLE".* This chapter presents the narratives of the interviews conducted with the subjects involved, the Northeasterners and/or their descendants, relatives who cut rubber in Acre's rubber plantations. It uses the DSC (Discourse of the Collective Subject) methodology to analyse the collective statements, experiences of these subjects. In this chapter, the migrants' narratives have a double representation: quantitative (with numerical expression) and qualitative (in the form of a discourse). This way of analysing was due to its effectiveness in processing and expressing collective responses, and also because it takes into account the collective societies of individuals. Subsequently, the discussions in this chapter were carried out with authors who discuss the subject.

Finally, there are the *"Final Considerations",* a revisiting of the entire discussion of the work and possible suggestions for further research. This Amazon of rubber tappers, northeasterners, rubber, rivers, forests, experiences was the scene of great moments that reconfigured this space, resulting in a new way of life, through the "heroes of the rubber tapper", the migrants from Ceará.

17

CHAPTER 1

GEOGRAPHICAL SPACE AND MIGRATION: THEORETICAL ASSUMPTIONS AND UNDERSTANDING IN GEOGRAPHICAL ANALYSIS

Image 01: Former rubber tappers Mr Divaldo Alves de Souza, 92, and his wife Oscarina Alves de Souza, 90.

Source: fieldwork, 2017

Photo: SILVA, M. L. S., 2017.

WHAT BELONGS TO ME

What is greater than me is part of me.
Rain fits in the sea, sand in the desert.
It's always been like that.

What's bigger than me I embrace like
God. Small things leak out. I cry for
them, maybe one night.

The next day the sun clears my soul.
I discover what is mine

forever. The dreams, the memories.
Our calm steps on the sand. The
laughter of my children, this belongs
to me.

(Bárbara Lia)

It's common to hear the expression Amazonian space, Brazilian space, home space, without first understanding it in its depth. Every day we talk about spaces without careful thought and we usually think only of mathematical space, that space that can be measured in metres and centimetres.

In geography, the concept of geographical space always varies according to its approach, through the various existing visions and definitions, often conflicting with each other, due to the epistemological and conceptual problems that geography has faced since its emergence or as a modern science. And despite major theoretical contributions in recent times, there are still theoretical disagreements about the definition of this object of study. Hence the importance of understanding these processes.

In this chapter we set out to emphasise some of the thinkers who contribute to the concept of geographical space, and then to discuss the concepts that guide the term migration, which we will highlight below.

1.1 Geographical Space

We'll begin our discussion of space with the ideas of Bollnow (2008) in his work *"Man and Space"*, which discusses how man relates to his space. Space is presented as a reference element for human existence.

When studying the individual's relationship with space, Bollnow emphasises two types of space: lived space and experienced space. The two have different meanings, since one carries sentimental values and the other does not. The lived space would be the one as a means of human life in which man lives in it and with it, without encompassing his psyche, just the space itself. Experienced space, on the other hand, would be in the subjective sense, taken as the experience lived by the subject. It would be understood as "the experience in space". For him, this would be a concrete space, since it is experienced by man, is endowed with meanings and has a realistic character, as he points out:

> [...] it can easily be taken in the subjective sense, as the way in which a space is experienced by a man, a space that, as such, is already there regardless of the way in which it becomes experienced, when the complement "experienced" refers only to the subjective placement that it superimposes on the space. Therefore, the term "experienced space" can easily be understood as "experience of space" in the sense of a simple psychological circumstance. On the other hand, the expression "lived space" is preferable when it expresses that it's not about anything psychic, but about space itself, since man lives in it and how he lives. It is space as a means of human life (BOLLNOW, 2008, p.16).

It can be inferred from the author's speech that the lived space would be the one that brings the experience lived by the subject, full of meanings and that generates a world reference space, the basis for the exploration of other places, it is a protective place, and it will interfere in their subsistence and in their understanding of themselves and the world. In other words, we must study space as a reference element for human existence.

Bollnow (2008) reveals a space of things, of movement, in evolution, free and extensive. For him, this would be "concrete space", because this is where life develops. According to him, the space experienced is not something of a spiritual nature, because it is not only experienced, but also imagined and conceived. It is something concrete, real, because that is where life happens. In this way, he warns us against thinking of this space only as something measurable, geometric, since we live and act in it.

However, it is not our idea here to discard the great legacy of Cartesian geometry, conceived in an abstract way, a fundamental concept for rational knowledge and of countless achievements such as geoprocessing and spatial navigation itself. The idea is to show, through Bollnow (2008), how man relates to his space, which is felt by our bodies and perceived by our minds because, for him, the space experienced would be richer and at the same time more complex.

This complexity leads us to think of the Amazonian space as a support for human life, endowed with meaning for the different peoples who live there. Being able to show a space endowed with feelings is our challenge here, especially for those who came from a place with climatic characteristics so different to the Amazon, the migrants from Ceará during the Second Rubber Economy, who arrived here, adopting this space as their own, taking root, adapting, adding new sentimental values through their daily lives and experiences within the rubber plantations.

This is why space is seen as a mechanism for the development and organisation of human life, always on the move, a space that is experienced and lived in a heterogeneous way, because that's where things happen. In Bollnow's ideas, the ground provides support for life, because man needs it to move around, it is his source of survival. The author's words reflect the example of the migrants from Ceará who used the rivers, forests and wild animals within the rubber plantations as a source of survival for their families. Having learnt to locate themselves in the dense and challenging Amazon rainforest from the first glance they took, through their daily experiences, they made this space their home.

Starting from the idea of space as "home", Bollnow reveals that man's relationship with space takes place in the action of inhabiting. But this "dwelling" action aims to build a space of reference. Thus, he compares this space of habitation with the house, the dwelling.

According to him, space is characterised by the parts of a house, because it deals with the concrete necessity of life, as follows: "But in losing a centre of the world, considering objective existence, man's life still remains referred to such a centre. It is the place where, in his world, he 'dwells', where he is 'at home' and where he can always 'return home'" (p. 134).

From this perspective, we have a space that is compared to a house, full of meaning for human beings, because it is a form of rootedness, belonging and intimacy. An experienced space, characterised by dwelling, a place of life and a fixed location, a symbolic universe of the human being. It serves the function of protection, cosiness, comfort and the essence of being well. It can also be a mythical space, felt and experienced because of its sacred character. It reflects the image of the world. Thus, this space, which he compares to a house, would be the "centre of the world" because it is full of images, deep meanings of the world, he says. Its walls serve to delimit its territories (BOLLNOW, 2008).

We realise that we have different spaces, endowed with a symbolic representation that is the result of the spatiality of human life, a space built through the relationship between man and his action of "inhabiting", a reference space, the "house", his dwelling, lived in, full of life, which in addition to its sentimental values also has the function of covering and protecting. That's why this inhabited space transcends geometric space, as it takes on sentimental, human qualities. Furthermore, Bollnow (2008) adds that this living space also gives the subject the freedom, through the "door", to set off on a journey in search of their survival and encounters with other subjects, in other words, it does not exempt them from appropriating new spaces to be explored for their survival and self-care.

An example of this is the migrants from Ceará who, already having a space of their own in their own land, opted, in a way, for this "freedom", to come to the Amazon, leaving their home space, full of meanings, mythical or sacred, to venture through the "door" that was opened to them, the call to "travel to the Amazon", to the "land of plenty", as they were advertised in the newspapers of the time (MARTINELLO, 2004). This corroborates the idea that man can carry his own space wherever he goes, because when he moves through space he takes his experienced space with him, his concrete space

Continuing with the search for definitions of "space", we'll turn to the ideas of geographer Tuan (1983) who, in an attempt to understand human action on space, distinguishes between space and place. The author emphasises that although these two terms are familiar and indicate common experiences, they differ and need to be distinguished. Thus, he leads us to the understanding that: we live in space (freedom) and place (security); we are attached to the latter, but we desire the former. For Tuan, spaces are demarcated and defended against invaders, while places are centres to which we attribute value and where biological needs are satisfied.

In the relationship between space and place, the meaning of space is often confused with that of place. It is with this in mind that Tuan goes further and emphasises the difference between space and place, as quoted below:

> "Space is more abstract than place. What begins as undifferentiated space becomes place as we get to know it better and endow it with value. [...] The ideas of "space" and "place" cannot be defined without each other. From the security and stability of place we are aware of the spaciousness, freedom and threat of space, and vice versa. Furthermore, if we think of space as something that allows movement, then place is a pause; every pause in movement makes it possible for location to become place. (TUAN, YI-FU. 1983, p.06)

According to this understanding, space (as an abstract) is transformed into a place to the extent that it is endowed with meanings, feelings acquired through accumulated experiences, in other words, the same vision that Bollnow has of "lived space".

This is why space is more abstract than place, since it doesn't carry these subjectivities. Tuan (1983) states that the perception of space will depend on the quality of the senses, the mentality and the capacity of the human mind to extrapolate beyond the perceived data, because such spaces are in the continuum of experience. This space is transformed into "place" as it acquires definition and meaning, because place is this world of organised meanings.

In this way we have: a space as a territory free of the senses. We could cite as an example an environment unknown to a subject, who has no affection for it whatsoever. As soon as this territory is filled with meaning, with subjectivity, it becomes a place. Like the example of the "house" in Bollnow (2008), this would be the "place" for Tuan (1983), because there is an affective link between the person and the place, or what he already called "*topophilia*" in his earlier work *"Topophilia: A study of the perception, attitudes and values of the environment"*, revealing what a place can symbolise to someone (TUAN, 1980). Adding another example, it would be the migrants from Ceará who, on arriving in the Amazonian "space", an unknown environment, without having any kind of affection, over time, this space was transformed into a "place" through the values and meanings acquired by it.

Based on these analyses, we can see that man transforms spaces into places through the feelings and sensitivities embedded in them. But this production of places, according to Tuan (1983), would only happen through their experiences, through direct or indirect human experiences that produce senses, sensations and notions about reality. This is why the hinterland is similar to that of the migrants mentioned, with such identification with their place when it comes to remembering their own in Benchimol (2010)[1] . The stories show that objects and people are transformed into places due to a variety of sensations such as safety, revealing the link between their past and the sense of "place". Furthermore, this reflection can help us understand the reasons why we appreciate certain places over others.

Also contributing to concepts of space are the ideas of Eric Dardel (2011), who made a major contribution to humanist geography and influenced geographers such as Yi-Fu Tuan. In his work "Man and Land", Eric Dardel shows us that the development of geographical science from the 19th century onwards was characteristic of the Western world. The debate focussed specifically on the *earth, space and matter.* And in order to understand the geography of the world, it is necessary to valorise the horizon of human life. Given this context, he characterises and conceptualises different types of space:

In *Geometric Space,* geography, in the etymological sense of the word, corresponds to the description of the earth. For this author, such descriptions of the environment are necessary to emphasise geographical expressions, since geographical discussions are constituted through the

---

[1] Although I understand the author's colonialist stance, I would like to emphasise here that the choice to include him in this dissertation is only as a source of data for the countless accounts of people from Ceará who came to the Amazon during the Second World War.

"world of the imagination", allowing more freedom for theoretical formulations that are conceived in the spirit and by its material world, laden with symbols. This ultimately allows the subject to dream, to the extent that this phenomenon is configured as a geography of the interior, where man in relation to the land can travel other paths in search of new environments, loaded with symbolic representativeness.

Material space is conceived as if it were a closed system in itself, and geographical space itself affects man within his social reality. Man is a producer of spaces, because at every moment we perceive a space in movement and under construction, so it is existential. Its construction will depend on the experiences gained in the world.

In *Teluric Space,* geographical space goes beyond the earth's surface, because geographical space cannot be conceived solely through perception acquired through experience, which is loaded with symbols. This space is permeated by images, dreams, feelings and sensations. This telluric space is fundamentally open to man, but we mustn't think of it as an inert space, it must be in motion just like the earth.

In the *Aquatic Space,* water takes on an extremely important dimension in the geographical space, as it sets the space in motion. The liquid that characterises this space is configured as a joyful element that transmits an inner peace of such repercussion. Man has many needs that go beyond the limits imposed by the world of water. For man, water has such significance that without its presence, the sea is a mere element in nature.

In *Airspace,* the geographical space that corresponds to the atmosphere, where the rain formed in the clouds is found. These aspects give meaning to man's life, but they are not always presented to him passively.

And finally, the *Built Space,* the one built by human beings, their living space, which, with their needs, will give it a new configuration, transforming the natural space. When these places are transformed, there are consequently ruptures, an established uprooting of man from the land, according to Dardel.

Given this, Dardel's geographical space is made up of different spaces and different horizons. A concrete and interested relationship between man and the land, the foundation of what he already called geographicity, a fundamental concept in his thinking. This concept expresses human relationships with spaces and places. It would be the constant complicity between man and his surroundings with his experiences, his lived world. These can be both positive and pleasant experiences and negative, unpleasant and repulsive experiences. Geography would include these good and bad encounters with environments (DARDEL, 1952).

In this way, Dardel has a world of everyday existence, which brings together dimensions of knowledge, action and affectivity, made up of differentiated spaces that reveal a set of signs. In this sense, human experience becomes indispensable because it displays a geographical reality. Therefore, this geographical reality will only make sense if man is connected to the land, because it is a space to be built and lived in by these subjects, a living space (living space), in which its structure would be configured through the creation of a kind of life that can meet their needs and establish relationships between subjects in their places.

Once again we have a space based on sensations, on human experiences, on the subjectivity that man can acquire in each place he inhabits and in the relationships between them. It can be seen that Dardel's (2011) geographical space is endowed with sentimental values, full of symbols, where life experience is a key factor, just like Bollnow's (2008) "lived space" and Tuan's (1983) "place". The intention of approaching the concepts of these thinkers was to understand the meaning of geographical space through the lens of the migratory process by Ceará residents in the Amazon region. This is the basis for reflecting on how these migrants approach and present this space.

In the migrants' narratives (presented in the next chapters of this work), we find a connection with the ideas of these authors through their various memories of their time in the rubber plantations and we see how they complement each other with these concepts. Dardel's spaces appear several times in his accounts, spanning the geometric, material, telluric, aerial, aquatic and constructed through his description, for example, of how they used this space to survive, using rivers and forests as a source of food, as a means of getting around; the aerial space as a way of guiding themselves through intense and dense forests or also the space they built when they arrived at each new place to cut rubber. Dardel's "geography", Bollnow's "lived space" and Tuan's "place" are also perceptible, as they often reminisce emotionally, full of nostalgia for the places where they have lived and worked. They show themselves to be subjects loaded with symbols and an intense intimacy with nature, through their experiences in the Amazonian environment. This space has an intense connection with the action of inhabiting.

However, before we can really reflect on what these people have to say, it is important to first introduce the displacements made by these people, migrations, an ancient phenomenon that can occur in any known location since the dawn of time. That's why we're going to look at the concepts that guide the word migration, which we'll cover in the next few topics, highlighting the "whys" of why they happen and also discussing the two periods that occurred in the Amazon for Ceará during the rubber extraction activity.

1.2 The meaning of migration. Why do they occur?

There are many challenges faced by scholars when it comes to defining the concept of migration. Over time, this subject has been discussed from many different perspectives. With this in mind, this topic will discuss the theoretical frameworks that encompass migration studies.

According to Becker (1997), the term migration is defined as the spatial mobility of the population, reflecting changes in relations between people and between people and their physical environment. And the displacement of these populations, according to the author, changes with each new world political order, in line with the economic order. Population groups set off in search of new territories, with a view to the possibility of land and a labour market, or simply wander in search of tasks for their subsistence. To summarise, migration is the movement of people from one place to another in search of better opportunities.

As far as studies of migratory phenomena are concerned, there are several theoretical strands. The first of these is the neoclassical approach. Until the 1970s, migration studies were considered from this perspective. Here, the decision to migrate is simply a personal, individualistic and sovereign decision on the part of the individual, without the interference of external factors. People migrate because they want to maximise their needs. From this perspective, all relations of domination in society are ignored (BECKER, 1997). The unit of analysis is only the individual and their natural propensity to move is just an assumption (SALIM, 1992).

This conception provides us with a reductive model of social reality, since analysis was carried out from an individualised perspective and exogenous forces were discarded. It was a limited vision that became ahistorical and supposedly apolitical.

But from the mid-1970s onwards, the phenomenon of migration was reconsidered from a neo-Marxist perspective, and migration came to be seen as mobility forced by the needs of capital (BECKER, 1997). Here, the migrant becomes a commodity on the move, seeking the maximum return on his investment at a given point in space (MONDARDO, 2009), he is a bearer of labour, and of interest in the processes of economic development (PÓVOA NETO, 1997).

Gaudemar (1977) conceived this mobility of labour as an element in the game of capitalism. The worker would take part in an economic game as a simple pawn on the board. The author criticises this system by saying that men are just instruments and their labour power is no more than a continuous mobility, moving only at the service of the machine and the capital that owns them. In other words, it would be the commodification of man through the mobility forced upon him by capitalism.

Along the same lines, in the context of internal migration, Singer (1998) conceptualised migration as a historically conditioned social phenomenon resulting from a global process of change. For this author, the first step towards understanding is to find the limits of the historical configuration that give meaning to a given migratory flow. For this reason, Singer lists two factors for migration to occur: the *expulsion factor* and the *attraction factor*. According to Singer, there are two types of expulsion factor: "change factors" - those where capitalist relations of production are introduced, leading to the expropriation of rural workers; and "stagnation factors" - those related to the growing demographic pressure on subsistence farmland, which is limited by the monopolisation of large landowners. The factors of attraction, on the other hand, were the demand for labour generated by industrial companies and the expansion of services.

Singer's ideas show that the factors of expulsion and attraction are preponderant. This means that for migration to take place, the place of departure must have bad obstacles that limit the individual's labour. The place to which they are going must be so attractive that they choose to migrate. These factors are caused by external forces that dictate the rules for these migrants.

The third theoretical strand is the historical-structural concept. In this, the analysis is rooted in the "soil", which seeks to explain the different economic levels of the territory, with the emphasis no longer on the sovereign decision of the individual, as mentioned above, but rather on the analysis of social groups and classes that suffer the force of social and economic structures, as we see: "migration redistributes the labour force according to the specific needs of the accumulation process, in concrete historical contexts". (MONDARDO, 2009, p.127).

The focus of this theoretical framework is redirected towards the contradictions within the social relations of production, the development of productive forces and the underlying mechanisms of domination. However, according to Póvoa Neto (1997), there are problems in reconciling the macro and micro levels, since migration is a social phenomenon whose determinants and consequences refer to other historically determined social phenomena and are related to processes of structural change. For Salim (1992, p. 125), "the historical-structural approach emphasises, above all, the view of the structure as a whole (...) the different movements of the population are explained by changes in the structure of production". He also warns against the "predominant tendency to measure migration by its economic aspect - modes of production, production relations, exploitation mechanisms, etc. - without relating it to other important macro-social processes, such as those of a social and cultural nature" (p. 126), because in this way migration only redistributes the labour force according to the specific needs of the accumulation process, in concrete historical contexts.

To summarise, the historical-structural conception is understood through the lens of social processes and not migrants, as in previous conceptions. Its focus is on the structural situation of society that results in a historical formation. That's why the point of analysis is these processes.

We also have "post-modernity" migration studies, where we find new elements incorporated into these migratory analyses. Mondardo (2007) mainly emphasises "cultural" elements, the relationship between the "self and the other", "psychological" elements, "identities", etc. For the author, this has to do with a "diversifying" perspective, "multiple interpretations" for a multitude of phenomena linked to migration.

Another way of understanding the migration phenomenon for this author is social networks. According to him, these networks build and structure opportunities, opening and closing opportunities. Migrants' networks of relationships influence their integration into the labour market of the receiving society. In this way, Mondardo emphasises that post-modernity offers a new reading, a new experience of the world, directly linked to the new technological paradigms that shake up the old certainties and the old ties between society and space. That's why the best way to understand subjective, structural and conjunctural elements would be to "blend" and/or multidimensionalise cultural and economic elements in the approach to migration. This would be done through analyses that recover the "everyday", the "place", the "subject", the "identity", the "representation of the self and the other", etc. This would reveal the complex world of migration and demonstrate that migration goes beyond the reality modelled by classical theories. For the author, this analysis would be a challenge in grasping reality and/or the phenomenon.

Another author who analyses this issue is Menezes (2007), who points to cultural elements, through loss, identity and collective consciousness. According to him, migration is always a loss. You lose your territorial reference, your identity, your cultural values and the people you know, which is what Bhabha (1998, p. 241) emphasises: "the transnational dimension of cultural transformation - migration, diaspora, displacement, relocation - makes the process of cultural translation a complex form of signification".

From this perspective, migration studies in post-modern times are diversifying in the face of the complexity of a globalised world, with different perspectives due to the immensity of events to which migration is linked.

Each stage of these theoretical trunks reveals that they are impregnated with a historicity, with ideas that are current in each one. This leads us to think that each moment is unique and needs to be considered according to its needs, as it demands and dictates new reflections and interpretations.

In view of this, we have set ourselves the challenge of analysing the migrations from Ceará to the Amazon around 1942, which, due to the production of rubber and encouraged by the Brazilian government, led to what we know as the "battle of the rubber". At that time, the north-east of Brazil was experiencing a great drought. Therefore, the incentive to migrate was a way found by the Brazilian authorities to solve the problem of unemployment and underemployment in the Northeast, which had worsened with the drought of 1942 (SOUZA, 1980).

Analysing these theoretical frameworks in terms of structural history makes us understand that the northeast was the main target at this time. Our focus is precisely on understanding the contradictions of the social relations of production in the face of the dominant productive forces of the time. But before going directly into this subject, in the next chapter we'll look at the methodological part of this research.

CHAPTER 2

THE PHENOMENOLOGICAL METHOD: RETURNING "TO THE THINGS THEMSELVES"

In the development of any scientific research, the methodological part is crucial and in Geography it would be no different, since this stage will provide support and order in the execution of the work. However, due to the great thematic diversity in geography, every method carries its own ideologies and epistemological positions, given the object studied. The method would be a way of obtaining a result based on a theory.

This diversity is essential for adjusting the method to the phenomenon, as Morin (2005, p.36) points out when he says that "The method can only be constructed during the research, it can only emanate and formulate itself afterwards, at the moment when the term becomes a new starting point, this time endowed with method". In this sense, we have the contributions of Bachelard (1983, p. 122) when he says that "The method is truly a cunning of acquisition, a new stratagem, useful at the frontier of knowledge" where the scientific method is "one that seeks out danger [...] and doubt is at the front, not the back", so "it is not the object that designates rigour, but the method" In this context we therefore have the method as a set of rational procedures.

This chapter will characterise and discuss the choice of method used, as well as the techniques used in this research and their contributions to human geography.

2.1 The Phenomenological Method

The search for the method of this research was based on the understanding that it would reflect an objective reality through the apprehension of the speeches of migrants from Ceará. This is because we concluded that it is not the researcher who chooses the method, but the method that chooses them. The choice of method was a fundamental part of this research and, as Spósito (2004) points out, it is based on the premise that it is the intellectual and rational instrument that makes us understand the various methodological interconnections of the study:

> [...] the method cannot be approached from a disciplinary point of view, but as an intellectual and rational instrument that makes it possible for the researcher to grasp objective reality, when he or she intends to read reality and establish scientific truths for its interpretation (SPÓSITO, 2004, p. 23).

In recent decades, phenomenology has emerged as an important method for geography, as a phenomenon that takes into account the different experiences of subjects, anchored in the way they feel, in other words, the individual's perception and knowledge (GOMES, 1996).

Thus, for the memory and imaginary of migrants from Ceará, the phenomenological approach stands out, because it contemplates their perceptions, subjectivities, emotions and affectivities. It was through this method that we were able to exchange empathetic feelings, to put ourselves in the other person's shoes, to feel the other person's pain and understand them as if they were one. The phenomenological method was chosen because of the commitment to these sensations, to valuing the sensitivities of these migrants, listening to each particularity. In fact, it was a bet on the imaginary donor consciousness, what Husserl (1989) states as: the supreme source of all rational affirmations.

The study of the phenomenological method is led by Edmund Gustav Albert Husserl, a mathematician and philosopher who is considered the "father" of this method and of the understanding of the human person, although he was not the originator of the term, but it was only through his works that this method was consolidated. He sought a method of grounding science, which established philosophy as a science of rigour, bringing this meaning to phenomenology. Thus, according to its etymology, the word phenomenology means "the study or science of phenomena". This phenomenology has the vocation of being *prima philosophia,* being the radicalisation of Cartesian thought. For this philosopher, reflection would be the authentic path of philosophical activity, which starts from the "I", from the experiences of the ego. Therefore, the method would consist of accessing the field of consciousness and then subjecting it to analysis (ZILLES, 2008). This meant a purely descriptive study of the phenomenon as it appeared in the light of experience. Japiassu and Marcondes (2001) provide more details on this:

> The phenomenological project is defined as a "return to things themselves", that is, to phenomena, to what appears to consciousness, which is given as its intentional object. The concept of intentionality occupies a central place in phenomenology, defining consciousness itself as intentional, as turned towards the world: "all consciousness is consciousness of something" (Husserl). In this way, phenomenology aims both to combat empiricism and psychologism and to overcome the traditional opposition between realism and idealism. Phenomenology can be considered one of the main philosophical currents of this century [20th century], especially in Germany and France, having strongly influenced the thinking of Heidegger and the existentialism of Sartre, and giving rise to important developments in the work of authors such as Merleau-Ponty and Ricouer. (JAPIASSU; MARCONDES, 2001, p. 101)

Husserl, in combating the adoption of empiricism and psychologism as the foundation of science and philosophy, sought to consolidate his phenomenology as a method. This is how he conceptualises it:

> It is the science of cognitive phenomena in this double sense: the science of knowledge as phenomena, manifestations, acts of consciousness in which these and those objectalities are displayed, become conscious, passively or actively; and, on the other hand, the science of these objectalities insofar as they display themselves in this way. The word 'phenomenon' has two meanings by virtue of the essential correlation between appearing and what appears. It effectively means what appears and yet it is preferably used for the appearance itself, for the subjective phenomenon. (HUSSERL, 1989, p. 34-35)

The phenomenological method proposed by Edmund Husserl aims to establish a secure basis, free of presuppositions, for all sciences. For him, positive certainties permeate the discourse of the empirical sciences, which he calls "naive". He believed that the supreme source of all rational affirmations is "imaginary donor consciousness". For this reason, phenomenology would not be concerned with the unknown behind the phenomenon, but with the given, without wanting to decide whether this given is a reality or an appearance: whatever it is, the thing is there (GIL, 1994).

The attempt at a "direct description of our experience as it is", and without any difference to its psychological genesis and the causal explanations that the scientist, historian or sociologist can provide for it, "is what characterises the phenomenological method". In this way, the method goes against the "truths" of rationalist science, presenting other forms of knowledge that are based on perception, worldly experience and the process of subjectivisation

considering perception, the lived world and subjectivity (PEREIRA and CORREIA, 2010). It is, in fact, the recovery of the experience "as it is", that is, the experience lived in space and time, allowing us to understand how the participants live, perceive, think and feel about their experiences. This is why the choice of this method for analysing the migration process from Ceará and describing the lives and experiences of these people in the new Amazonian space is so fascinating.

This method is neither deductive nor empirical; its interest is in showing and clarifying what is given. Nor does it seek explanation through laws, nor does it deduce from principles, but it immediately considers what is present to consciousness, the object. Consequently, its tendency is totally orientated towards its objective, its interest is not in the subjective concept, nor in the activity of the subject, but in what is known, doubted, loved or hated (GIL, 1994).

Phenomenology makes it possible to understand what it means to be a human person. In light of this, Bello (2014) reflects on this meaning, emphasising that what differentiates the human person from other beings is our awareness of the questions we ask ourselves as living beings. Drawing a comparison between us and a plant, he points out that a plant, despite having a body that is tactile, a soul (which turns to the sun in search of light for its photosynthesis), still doesn't have a spirit that asks what it is, what it represents, what it feels. So that would be the main difference, because we are questioners of what we are and the actions we take in the world around us. That's why the task of phenomenology is:

> The task of phenomenology, or rather the field of its tasks and investigations, is not as trivial as if we only had to look, simply open our eyes. Already in the earliest and simplest, in the most intimate forms of knowledge, the greatest difficulties are posed for pure analysis and pure consideration of essences; it is easy to talk in general about correlation, but very difficult to elucidate how a cognitive object is *constituted* in knowledge. And the task is now, within the realm of pure evidence or the given in itself, to *trace all the forms of the given and all the correlations and to* exercise enlightening analysis over all of them. (HUSSERL, 1989, p. 33)

In this sense, it can be seen that the role of phenomenology is to enable a more intrinsic understanding of what it is to be human and what motivates us to survive, i.e. not with a superficial look at the object, but an exchange of empathetic feelings that make it possible to understand what it is to actually be a human being. Not only does it understand what it is to be a human being, but it helps us to practice putting ourselves in the other person's shoes, being empathetic, being sensitive, but also understanding them as if they were one, because, unlike other living beings, human beings can see that another person has the same physiological needs as them, and this is what characterises them as rational beings.

According to Zilles (2008), the German philosopher Immanuel Kant already treated phenomenology as the study of a set of phenomena or appearances that manifest themselves in time and space. He used the term to characterise the "propaedeutic discipline" that must precede metaphysics and also to explain what there is of sensible intuition in objectivity and what does not appear, but is purely thought: the in itself. In these terms, there is a concern to emphasise the essence.

In this way, existence is re-established, insofar as the tangible has always existed "there", in a form prior to thought. This intellectual, spatio-temporal abstraction of the "lived" world materialises when describing experience as it occurs (MERLEAU-PONTY, 1999). Merleau-Ponty goes on to criticise modern thought, because in his view it inverts common sense, since the preoccupation with truth and experience does not allow one to honestly stick to the clear or simple ideas that common sense clings to, because they bring it peace of mind.

Phenomenology comes in as an attempt at a direct description of our experience as it is, because the meaning of phenomenology lies in ourselves, in the lived world and in the space-time relationship. In making these criticisms, Merleau-Ponty emphasises that rationalist thinking, based on objectivity, disregards the subject of perception because the subject encounters a world that is totally ready-made. However, a being who perceives becomes part of things and is unable to detach himself from them. He ends up producing a perceptual impression of phenomena and can describe things from a distant place, even if he hasn't been there.

It can be concluded that the phenomenological method makes the researcher and the researched interact in activities with empathy, so that what is lived and experienced by the researched becomes lived and experienced by the researcher, resulting in a true and complete understanding of the other. For this reason, Merleau-Ponty (1999), unlike modern science, which makes a dichotomy between the objective and the subjective, or in other words, between subject and object, proposes a break with these dichotomies in order to bring the subject and the object closer together, because the subject, as a researcher, needs to turn to himself, which would be a *return to the things themselves,* and face these things as they really are. In this new approach, Merleau-Ponty becomes a reference for geographical studies based on a humanistic and perceptive vision.

In this sense, the phenomenological method for analysing the memory and imaginary of migrants from the Northeast at the time of the Second World War could not have been more satisfactory, since through it it was possible to fully experience what these migrants went through, dreamt of and experienced when they came from the Northeast to the Amazon and kept themselves alive and strong in the face of the work in the rubber plantations, in an inhospitable region, with a climate different from their own, difficult to access, along with all kinds of deprivations inherent to the place. The method chosen during the interview allowed for an exchange of empathetic feelings, so as to be able to put oneself in the other person's shoes, to be able to feel the other person's pain and understand them as if they were one. It was by focusing on these sensations, on valuing these sensitivities, listening to each particularity, that phenomenology was chosen.

2.2 Technical Processes

In this section we will discuss the methodological tools used to understand our object of study. We chose to use four fundamental techniques: Bibliographical Research; the Snowball Technique; Oral Memory and the Collective Subject Discourse Technique - CSD.

The techniques chosen range from the qualitative approach, as it emphasises the perceptions that are unique and inherent to human beings and that can only be achieved through this approach. It is seen here, according to Bogdan's (1994) ideas, as an examination of the world in which nothing is trivial, but in which we can establish a more enlightening understanding of an object of study. In other words, there is a dynamic relationship between the real world and the subject, an inseparable link between the objective world and the subject's subjectivity. However, we set ourselves a greater challenge, which was to also involve this research in a quantitative way, or rather a qualitative-quantitative way, through the use of the Discourse of the Collective Subject - DSC, which achieves this double representativeness of the collective opinions and ideas of the interviews. Here, each distinct collective opinion is presented in the form of a discourse while the quantitative representativeness arises from the fact that this discourse has a numerical expression that indicates how many statements were needed to make up each CSD. But it should be emphasised that the DCS is about giving visibility and representativeness to the facts, to the documents. This point will be discussed in more detail in the following sections.

2.2.1 Bibliographical research

Bibliographical research was fundamental to this study, as it allowed us to delve deeper into the subject from the perspective and perceptions of other researchers. Although this is not an unprecedented piece of research, we understand that it is always important and necessary to read and reflect, and to get hold of a wide range of information to add to the initial ideas. Gil (2012) discusses bibliographical research:

> Bibliographical research is carried out using material that has already been prepared, consisting mainly of books and scientific articles. Although almost all studies require some kind of work of this nature, some research is carried out exclusively from bibliographic sources (GIL, 2012 p. 50).

This shows the importance of bibliographic sources for the production of scientific work, which serve to add value to the subject. However, despite being an inexhaustible source of knowledge and discussion, it is essential that these bibliographic sources are reliable and of a scientific nature in order to avoid spreading possible misinformation. Gil (2012) corroborates this:

> Secondary sources often present data that has been collected or processed in the wrong way. Thus, a study based on these sources will tend to reproduce or even amplify their errors (GIL, 2012, p. 51).

Having explained these ideas, we can see the importance of this technique for reflecting on migration from Ceará to the Amazon. The sources obtained were theses, books, electronic magazines, scientific articles, newspapers and newsletters. The stages were as follows:

J Preliminary bibliographic survey of the subject in libraries, websites, blogs and electronic magazines. We also made use of works discussed and debated during the subjects taken during the Master's programme;

J Readings and summaries of these and logical organisation of the text (provisional writing);

J Writing the text.

These stages were essential and the readings carried out led to a greater apprehension of the reality studied. The theoretical study of the concept of space and migration, more specifically migration to the Amazon/Acre until the creation of this new Territory, was based on the ideas of other authors such as Becker (1997), Mondardo (2009), Singer (1998), Peliano (1990), Martinello (2004), Pontes (2014), Benchimol (2011), Nascimento Silva (2000), Medeiros Filho and Souza (1984) and Oliveira (1985). Their contributions only endorse how great and significant migration from Ceará was in the Amazon region. This signifies the importance of these studies, which is decisive when it comes to contrasting them with the vision of the interviewees in this research, who experienced this situation in a particular way "from the inside out" of this new constructed space. Understanding this space was also indispensable through the ideas of geographical space presented by Otto Bollnow (2008), Eric Dardel (2011), Tuan, Yi-Fu (2983), given the importance for human beings of this constructed and lived space, as well as the need to keep man and nature always interconnected. All these steps were necessary support for the understanding, logic and development of the research.

2.2.2The   Oral Source

In this study, we also opted for oral sources, which over the centuries have proved to be a source of preservation and dissemination of knowledge for science in general.

(GONÇALVES and LISBOA, 2007). It predates drawing and writing, cited in Thompson's (1992) studies when he says that it is as old as history itself, as it was the first kind of history.

It was because of the importance of knowing where and how this oral source would be stored that the tape recorder emerged, a resource that would greatly help in this type of research. Alberti (2005) states that the technique of recording a story became more widespread in the middle of the 20th century, after the invention of the tape recorder, which consists of conducting recorded interviews with individuals who took part in or witnessed events and circumstances in the past and present. Therefore, it basically consists of conducting recorded interviews with people who can testify to a phenomenon, way of life, among other aspects. In this research, the accounts collected from the selected participants were through this technique.

The oral sources collected (recorded) in this research were fundamental because they were migrants from Ceará, the "rubber soldiers", recruited during the 1940s to cut rubber in the Amazon/Acre. Also the children and/or grandchildren of these soldiers who worked in rubber extraction in Acre's rubber plantations and experienced the "rubber tapping process" and, finally, the wives of rubber tappers who worked in rubber extraction in Acre's rubber plantations. A total of twelve interviews were carried out in the municipalities of Mâncio Lima, Tarauacá and Rio Branco.

Listening to these stories was like going back in time, making this time "present", something alive, which is actually our endeavour here. It was a way of expressing a past, of bringing this past to the present. Poulet (1992) states that it is thanks to this memory that time is not lost, which reflects the phenomenological process in this research.

All awareness of past events would serve, through these recovered memories, to distinguish yesterday from today, confirming a lived past (LOWENTHAL, 1981). Hence the importance of recording the memory and imagination of the rubber soldiers and their descendants who still live in the state of Acre.

For this reason, the choice to work with oral accounts aims, above all, to give space to the history of these people who were mostly invisibilised, not valued, the basis of the migratory phenomenon in this region. However, it must be emphasised that during the various interviews, the act of reminiscing was not always a healthy or positive action for them; sometimes there were feelings of pain and suffering, alternating with longing and a certain nostalgia. It is precisely in listening to and describing these stories that we can narrow down the field of our enquiries, allowing us to describe the representations of these individuals who lived through history or, in some way, had contact with it.

However, in order to carry out this stage, the oral technique required activities before (searching for and selecting interviewees; drawing up questions) and after recording the testimonies (transcribing for analysis). The initial planning for the interviews and the definition of the participants was helped by my own family in conversations. In order to reach all the interviewees, it was necessary to use another technique, known as snowballing, which will be discussed in the next topic.

2.2.3 The Snowball Technique

This type of sampling is a form of non-probability sampling, which uses chains of reference and is also very useful for studying certain groups that are difficult to access (VINUTO, 2014), which is not our case, the people from Ceará and their descendants who live in the state of Acre. These people are still easy to find.

The way this technique was used was very simple and went as follows: for the kick-off, we looked for key informants, referred to here as "seeds", in this case represented by my own family, in order to locate some people with the profile needed for the research, within the population of the state of Acre. These "seeds" then helped to contact other people to be interviewed. These interviewees were then asked to indicate new contacts with the same historical characteristics, within their own personal network, and so on, growing like a snowball system. In this way, the sampling frame reached the desirable size.

Having defined the profile of the interviewees, the following steps were taken:

- First stage: Locating the interviewees in the municipalities of Acre, through acquaintances of the researcher's own family (who are of Northeastern descent), using the "snowball" technique - when they indicated another and so on;
- Second stage: Scheduling the most convenient day, place and time, respecting the preferences of each of the research subjects;
- Third stage: travelling to the interviewees' municipalities;
- Fourth stage: visit and realisation of the interview (recording).

During the interviews there were pre-selected questions to start (call) and lead the conversation, but care was also taken in this approach, making it non-systematic, trying to make these subjects as comfortable as possible, as in an informal conversation. That's why, as the conversation unfolded, it was no surprise that new and interesting subjects came up that were different to those usually told in historical books.

To ensure the safety of these accounts, and with the permission of the participants, all the interviews were recorded, stored on instruments (mobile phone recorders and later on a computer), and then transcribed in full. At the end of the interviews, photographic records were taken of all those who gave their permission, together with the researcher, as a way of saying thank you and revealing the beauty of the moment.

Two other interviews were also carried out with people who live in hard-to-reach places in the interior of Acre. In order to make initial contact, it was necessary to enlist the help of a resident of Rio Branco Acre, Mrs Maria de Lourdes Souza Silva (the researcher's mother), who volunteered to help in the process of identifying the interviewees.

At this stage, the interviews were recorded via mobile phone using a phone application called *"Call record"*. Once the audios were recorded, all the interviews were transcribed in full, taking care to respect and value the regional words and expressions spoken by the interviewees.

All the interviews were designed to reach the most distant points where these migrants found themselves in Acre. To do this, we opted for the three gateways, which are the three rivers that cut through the cities and through which these people entered in the 1940s: The Acre River (in the city of Rio Branco, far east of the state); Tarauacá River (in the municipality of Tarauacá, north of the state) and Juruá River (in the municipality of Mâncio Lima, far west of the state).

The use of the two aforementioned techniques was essential in this research, since bringing back the accounts of these subjects who, in a way, have been excluded and placed in anonymity, without the right to memory, is like making a voice that has been dormant for a long time speak.

Finally, we adopted the Collective Subject Discourse (CSD) technique, which, although recent, we believe to be innovative and useful for this study, shown in the next topic.

2.2.4 The Collective Subject Discourse Technique (CSD)

The Collective Subject Discourse (CSD) technique is based on qualitative and quantitative methods:

> The CSD presents the dual representativeness - qualitative and quantitative - of the collective opinions that emerge from the research: the representativeness is qualitative because in research with the CSD each distinct collective opinion is presented in the form of a discourse, which recovers the different contents and arguments that make up the given opinion on the social scale; but the representativeness of the opinion is also quantitative because this discourse also has a numerical expression (which indicates how many statements, out of the total, were needed to make up each CSD) and therefore statistical reliability, considering societies as collectives of individuals. (LEFEVRÈ and LEFEVRÈ, 2006 p 02)

According to Lefevrè & Lefevrè (2006), this methodology is a way of making the collective express itself directly and is widely used in qualitative research in the Human Sciences. It is an explicit proposal to reconstitute a collective, opinionated entity in the form of a subject speaking in the first person singular, in other words, speaking as if they were an individual, but conveying a representation with collective and amplified content. The CSD is therefore an elaboration of pieces of discourse with a similar meaning, brought together in a single discourse, in order to make the collective speak as if it were a single individual (LEFEVRÈ and LEFEVRÈ, 2003).

The technique tabulates and organises qualitative data in order to resolve one of the great impasses in that it makes it possible, through systematic and standardised procedures, to aggregate statements without reducing them to quantities. Basically, it analyses the verbal material collected from the surveys to compose one or more synthesis discourses, which are the Discourses of the Collective Subject (LEFEVRÈ and LEFEVRÈ, 2003).

One of the aims of this technique is to reduce the variability that is naturally present in discourses, in order to validate the knowledge that the author of the discourse represents in his speech. This is why the freedom of the interviewees to speak, think freely and argue must be valued. Structurally, it is organised using methodological figures, a set of speeches that show a collective thought or the group's representation of a given theme or issue (SALES & SOUZA, 2007).

In this way, the CSD achieves double representativeness, i.e. both qualitative and quantitative, of the collective opinions and ideas of the interviews. The representativeness is qualitative, because in research analysed using the CSD, each distinct collective opinion is presented in the form of a discourse representing an opinion on the social scale. Quantitative representativeness arises from the fact that this discourse has a numerical expression that indicates how many statements were needed to make up each CSD, represented in the proposed categories (LEFEVRÈ & LEFEVRÈ, 2006). According to these authors, in order to construct the CSD, it is necessary to take advantage of all the ideas present in the testimonies, and also to link the discourses narratively so that they present a coherent and clear sequential structure.

In this sense, the CSDs are produced using four elements:

Key Expression', literal extracts of parts of the statements that best represent the content of the speeches related to the objective of the question;

Central Idea', synthetic formulas with the essence of the discursive content, must be concise, respecting the content and meaning of the answer issued by the subject, identifying which idea is expressed;

Anchoring: here we find explicit linguistic traces and theories, concepts and ideologies that exist in society and culture, but which have not been internalised by individuals;

Discourse of the Collective Subject', is reconstructed with parts of individual discourses, as a sign of knowledge of the discourses themselves. The Key Expression of the Central Ideas or Anchors with the same meaning of thought present in the discourse. In short, it's as if it were a real person speaking, with collective thought as its content (LEFEVRE & LEFEVRE, 2003).

Thus, it is from the central ideas and their respective key expressions or anchors that various summary discourses are composed, the so-called Collective Subject Discourses. This systematisation will make it possible to retrieve the collective opinions gathered through individual interviews with open-ended questions, allowing thoughts to express themselves and, with this, letting the affectivity of social representations come to the surface.

In this way, the new tool represents a significant change in the quality, efficiency and scope of qualitative research, allowing us to get to know, with the safety of scientific procedures, in detail and in their natural form, the thoughts, representations, beliefs and values of every type and size of community, on every type of topic that concerns them (LEFEVRE & LEFEVRE, 2003).

The techniques used served as "ladders" in the process of this research. Each one helped in its own unique way. However, it is also important to be able to reflect on the processes that drove migration to the Amazon, pointing out the main reasons that generated an intense rush to the new space. This subject will be discussed in the next chapter.

CHAPTER 3

THE ELDORADO GRID: FROM THE TWO GREAT INTERNAL MIGRATIONS TO THE CONSTITUTION OF THE STATE OF ACRE

I saw a young man from the caatinga meditating

About the pain that haunts your life

So needy, so hungry, so suffering

Like others, the cemetery has silenced

The drought industry knows how to kill

Under the orders of the coronelista power

Latifundia of the capitalist sphere

Who are also the masters of power

Northeast Ethiopia now dreams of living

In the empire of the extractivist kingdom.

I saw a hopeful people migrate

To escape misery and poverty

Finding dignity and wealth

In the greenest place

The government made the people believe

On the real success of extractivism

Being a rubber soldier is "heroism"

In the illusory world of extraction

And the nation's sovereign forest

He went to serve imperialism's table.

(Francisco Marquelino Santana).

The settlement of the state of Acre by people from Ceará began at the end of the 19th century with the extraction of latex, through an intense wave of migration. In the 1940s, another migratory flow took place towards the Amazon. Thousands of people from Ceará, called up and recruited during the Second World War, once again went into the forest to cut rubber, driven by a common dream: to be able to make a fortune and change their social position in their region. These migrations were triggered by a set of factors organised, thought out and articulated by Brazilian bodies and, in accordance with international treaties, which resulted in the formation of a new place, a place called Acre, previously Bolivian land, which had been conquered at the cost of bloody battles.

In this chapter, we will discuss the internal and external factors that contributed to the formation of the territory of Acre, highlighting the two great migrations, the basis for the process of settlement and conquest of space by the "white man". It is essential to understand the global context in which Amazonia was situated, to analyse the dominant policies and the technological transformations that the world's industrial powers were undergoing at the time.

3.1 The first moment of migration from Ceará to Amazonia

For a long time, the Amazon region, known as inhospitable and unhealthy, didn't arouse much interest in non-indigenous people doing any kind of work. Only the various indigenous peoples who lived in it knew, knew and mastered this area well. The reasons that led to the reversal of this situation are inextricably linked, above all, to external conditions.

The Second Industrial Revolution, which began in the United States around 1870, saw metallurgy, electromechanics and petrochemistry as branches of industry, and electricity and oil as forms of energy. It was a time of another material civilisation. Metals (steel, the basis of everything) lead humanity to a state of geological civilisation and electricity and oil to a civilisation of energy. Oil gives rise to the explosion engine and establishes the petrochemical industry alongside the metallurgical industries with great expression, moving transport forward, where the motorway adds to the railway and navigation, a source of network integration with great speed and capacity for movement. At this point, the automobile industry came into its own, assuming the centre of gravity of the system, the symbolic image of the Second Industrial Revolution. (MOREIRA, 2000)As an example of this context, the diagram below (fig. the situation of the Amazon in relation to the technological transformations that the major world powers were experiencing and which required the search for sources of raw materials.

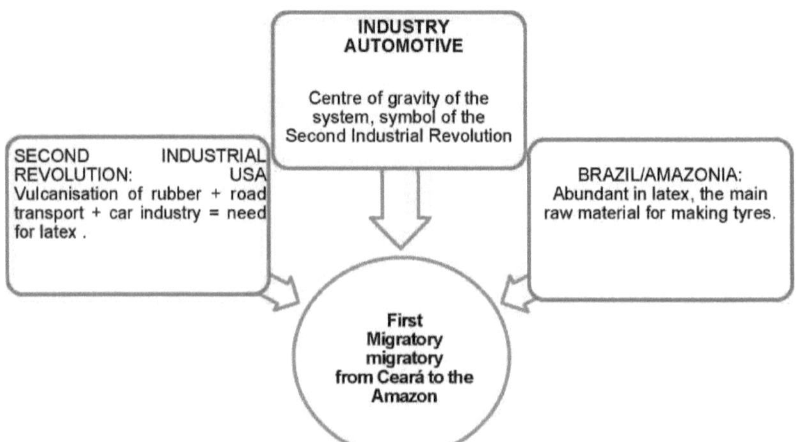

Figure 02: Internal and external context of the Amazon at the end of the 19th century

Org.: Prepared by SILVA, M. L. S., 2017.

It can be seen in this context that the new technological revolution, in this case the automobile industry, demanded the search for fields to supply its raw material, latex, which technological progress now required, with the discovery of the rubber vulcanisation process[2] . This prompted American industries in particular, but also European ones, to accelerate their search for latex, placing a new value on Amazonian rubber. Acre, a region that until then belonged to neighbouring Bolivia, was rich in this raw material (MARTINELLO, 2004).

About this Pontes (2014), describes:

> Technological progress in the chemical, steel and electrical industries, during the period known as the Second Industrial Revolution, accelerated the demand for rubber and transformed it from a simple "backwoods drug" into a stable product that was widely used on an industrial scale, especially in the North American and European industries. The growing car industry and the use of rubber tyres accelerated the search for areas producing latex, which became a key component in international industries. Thus, the stage was set for the exploitation of Amazonian/Acrean rubber (PONTES, 2014, p.02).

The technological modernity of the Second Industrial Revolution brought with it the need for latex for the car industry. It was precisely at this time that the eyes of the world turned to the Amazon region, rich in this resource that would make it a world export powerhouse, leading thousands of people to rush to this region. Benchimol (2011) points out that rubber reached export levels never seen before, given the importance of this product. He also gives figures for the increase in the use

---

[2] Vulcanisation is the operation by which bonds are created between the macro-molecules of an elastomer. In this way the elastomer, which at first appears as a weak, very plastic mass with no interesting mechanical properties, is transformed into a strong, resistant product with good elastic characteristics. When he discovered this technique, Goodyear applied it to natural rubber, using sulphur as the vulcanising agent (GUERREIRO, 2003).

of this raw material throughout the 19th century:

<blockquote>
From 1827, Amazonian rubber began to appear on the Amazonian export list with shipments of 3 tonnes. With the discovery of the vulcanisation process in 1839 and the increased use of this raw material, records rose to 1545 tonnes in the 1840-1844 quadrennium; in the 1875-1879 period it stood at 30,360 tonnes; in 1901 it would reach 30,241 tonnes and at the height of the boom, in 1912, it reached 42,286 tonnes, rates never reached before. [...] (BENCHIMOL, 2011, p.73)
</blockquote>

This situation became an attraction for thousands of people, mainly from the north-east of Brazil, to come to the Amazon. In this case, thousands from Ceará, coming from a region where, among other problems, there was strong social inequality and poor distribution of land and income. They arrived in search of better days and the longed-for Eldorado, an image given to them by a region of *"plenty, milk and mef"*.

Martinello (2004) explains that there were several factors that led to their migration, among them: the prejudice for toiling in the coffee plantations of the southeast; the illusions of rapid enrichment exposed by the rubber boom; the advertisements carried out by rubber tappers to attract this labour force; the government subsidies granted for transport; the ease of cabotage transport (merchant shipping) to the port of Belém; breaking the resistance of northeastern landlords to the departure of men; as well as the famous drought of 1877 that ravaged that region consequently forming population surpluses in the workplaces. In fact, the author reports that it was the drought that was the determining factor in the exodus of people from Ceará to the Amazon. In addition, the incentive provided by the national government was only for cheap labour, already accustomed to a rougher life. The propaganda about the Amazonian lands would solve the problem of the great drought that the north-east of Brazil was experiencing at the time. It was seen as certain and easy wealth.

It is estimated that from 1850 to 1900 the population of the Amazon valley increased tenfold. Benchimol (2011) highlights these figures. From 1877 to 1883 there were 19,910 migrants; in 1892, 13,593; and in the three-year period 1898/1900 there were 88,709 migrants, the peak of the population movement. Adding up the numbers, we would have 158,125, around 20 per cent of the Amazonian population at the time.

This rate of migration was so high in this period that, according to (NASCIMENTO SILVA, 2000, p.49), "it worried the large landowners of the Northeast, because it left the rural areas of the Northeast short of labour".

This astonishing migration had incentives from the Brazilian government behind it, including free tickets, so that they could move to this region. And even those who didn't want to were compelled, because the government even used police force to force them to migrate, and the drama was visible, as Medeiros Filho and Souza (1984, p.59) point out: "There were cases of shipments carried out by force in which the husband went north and the wife went south".

In Acre, these northeasterners began arriving in 1877 to cut rubber, discarding the existing indigenous labour force. Despite the fact that the native element contributed greatly with their customs, ways of life and ways of respecting the laws of nature, some rubber tappers viewed them with contempt, labelling them as lazy, naughty and tricky. However, these native customs were widely learnt by the "brabos" who arrived with no knowledge of the region. The large human contingent arriving in Acre was made up of men from the backlands of Ceará, Pernambuco, Paraíba and Rio Grande do Norte (MARTINELLO, 2004). Oliveira (1985) adds:

> [...] the population exodus to the Amazon occurred in the context of the expansion of the extractive company after 1850. While until then rubber extraction had been confined to the outskirts of Belém and then to the Xingu and Tapajós, between 1850 and 1870 it reached the province of Amazonas, reaching the Madeira and Purus rivers, bringing a first flow of migrants to the western region. This flow was already partly made up of northeasterners, in some famous expeditions that ended up putting down roots in the region of these two great rivers. These were groups of relatives and friends who embarked, mainly in Ceará, with their families and neighbours in search of enrichment in the rubber plantations that still had no owners. (OLIVEIRA, 1985, p.13)

According to Oliveira (1985), this intense flow to this region was due to the huge reserves of rubber plantations, resulting in a rush to what was considered the "great Eldorado", as shown in figure 03:

Figure 03: Acre's rubber plantations during the first migration

Left: Northeasterners in Amazonian rubber plantations during the first rubber economy. Right: Seringai Nova Floresta on Acre lands, owned by Mr Soares Hermanos, early 20th century.

Source: ALMA ACREANA, 2012, adapted by SILVA, 2017.

The images above, from figure 03, show some of the thousands of rubber tappers who arrived here, driven by the discourse of abundant land, promoted by those who articulately sought only their own interests and profits. It was the beginning of an economy that, although slight, would command the Amazonian space with such force as no other had done before.

The rubber plantation, in the first phase of the boom in the rubber economy, was a hostile world for these migrants who, without their families, without subsistence farming, since their only duty was to cut rubber, and their routine on the rubber roads took up all their time for other chores, were forced to buy and sell to their bosses at harsh prices. The rubber tapper had nothing of his own, no "producer status" (OLIVEIRA 1985)

This phase remained high, reaching its climax in 1912. But what seemed unimaginable happened next, starting with a slight drop in rubber production. The year 1912 was to be the peak and the end of Brazil's rubber export hegemony. The reasons for the surprising loss of market share were the emergence of Asian competitors, who, as a "masterstroke", now owned rubber plantations and exported more rubber. This was due to the secret collection and dispatch of the plant's coveted seeds. English botanists like Henry Wickh came to the region between the Tapajós and Madeira rivers as if they were orchid collectors. They collected more than 70,000 seeds without anyone knowing, and took the "orchids" back to their continent for experimentation. They were our own Amazonian seeds being biopirated, being grown in a different place, at a lower final cost than in Brazil, which now competes with and reduces the world's interest in Amazonian rubber. (MARTINELLO, 2004).

By 1920, Asian production had already surpassed Brazil's 10 times over. In the 1930s, Brazil would surprisingly contribute an inexpressible 2% of production. There were more than 800,000 Asian tonnes against 17,000 Brazilian tonnes (MARTINELLO, 2004). What's more, in the depths of the 1932 crisis, Brazilian exports reached just 6,244 tonnes (BENCHIMOL, 2010). The figures show the huge disparity in exports between the two countries, which meant that Brazil could no longer compete. It was the loss of Brazilian supremacy.

With the end of the first rubber boom, chaos ensued in the economies of Belém, Manaus and Rio Branco, which saw their finances fall to 1/3 of what they had been. For this reason, many rubber plantations were shut down with debts owed to the aviation houses[3] . At this point, a reverse migration began, the opposite of what had occurred previously. Without the income from latex extraction, some still managed to return to their states of origin, but many others could not, due to the countless difficulties they encountered, moving to and settling on the outskirts of the state capitals, and in their desperate search to escape the calamity, they found misery and unemployment there (MARTINELLO, 2004).

According to Oliveira (1985), Acre saw the decay of small towns and villages. Rubber plantations, both large and medium-sized, were abandoned by rubber tappers and seringalistas. Aviation companies took over countless rubber plantations. In the 1920s, Rio Branco's population of almost

---

[3] Located in Belém and Manaus, these aviation houses were commercial establishments that supplied the rubber plantations, receiving the rubber produced from them in return and then selling it abroad (SILVA, 2010).

20,000 fell to just over 15,000 by the end of 1930. The rubber tapper had been left to fend for himself, now a beggar.

3.2 The constitution of the Territory of Acre

The formation of Acre's territory took place in the context of the expansion of the extractive industry, when rubber became an indispensable raw material for the emerging consumer durables industry in Europe and the United States, mainly in the form of tyres. International interest in the Brazilian, Bolivian and Peruvian Amazon resulted in a series of border issues. The large number of migrants arriving in Acre was due to the fact that it was one of the main extractive areas and soon stood out among other regions of Brazil for its production at the beginning of the 20th century. But in order to understand how the process of Acre's formation took place, it is necessary to look at the antecedents and contributors to its history.

In the first division of the American continent between the Portuguese and the Spanish, the lands of present-day Acre belonged to Spain. This was legitimised by a document called *"Bula Inter Coetera"* in 1493, issued by Pope Alexander VI at the time. Portugal did not accept this and in 1494 signed a new agreement, the Treaty of Tordesillas. Three more agreements were signed between Portugal and Spain, namely: Treaty of Madrid (1750), Treaty of Prado (1761) and Treaty of Santo Ildefonso (1777). But it was in 1867 that a new agreement was signed on issues related to the Acre region, now between Brazil and Bolivia, both free from Iberian colonial rule, by the Treaty of Ayacucho, where the Acre region would officially be Bolivian, despite being recognised as being occupied by Brazilians (OLIVEIRA, 1985; PONTES, 2015).

At that time, the region belonging to Bolivia, which was rich in latex, was gradually penetrated by rubber tappers and rubber tappers. Oliveira (1985) describes that before 1870, Brazilians were already entering the area and after 1880, the formation of rubber plantations was greatly accelerated by the large waves of immigrants arriving from the Northeast. The wave of migrants made the rubber plantations thrive on the banks of the Acre, Purus, Juruá, Madeira, Tarauacá, Abunã, Laco and Beni rivers, among others. For rubber tappers or rubber tappers, it didn't matter whose land it was, it was all forest, land to exploit, to expand.

However, the region was rightfully Bolivia's and it had recently established a free trade agreement between the two countries. According to Mello (1990), on 27 March 1867, Brazil and Bolivia strengthened their trade relations through an agreement called the "Treaty of Ayacucho[4] " or

---

[4] Signed between Brazil and Bolivia on 27 March and sealed on 23 November 1867, in the Bolivian city of La Paz, with the then Bolivian president, General Mariano Melgarejo and the Brazilian Emperor, Dom Pedro II, the Treaty is part of a series of geopolitical and military agreements related to the territorial demarcation of the regions belonging to the current Brazilian state. As a result, traffic

the "Treaty of Friendship". This was one of the most liberal treaties concluded by Brazilian diplomacy. It prescribed a revision of the geopolitical boundaries established by previous treaties, from which Bolivia's border reached the regions of the current states of Acre, Rondônia and part of Amazonas, going as far as the Madeira River, near the current city of Humaitá, a municipality in the state of Amazonas, on the Jamarí River. The new treaty resulted in the retreat of Bolivia's borders, allowing for the expansion of Brazil's frontiers.

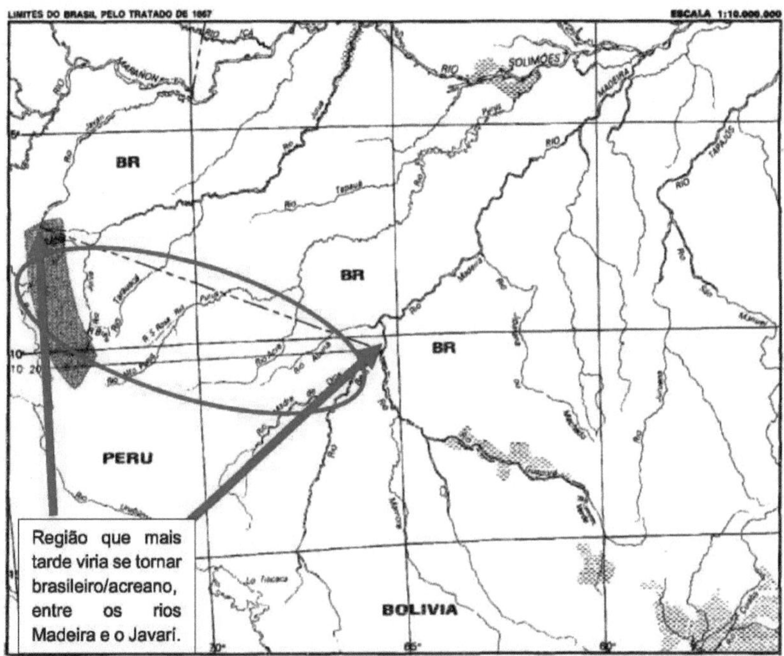

Figure 04: Map of the boundaries of Brazil by the Treaty of Ayacucho 1867
Source: MELLO, 1990 organised by SILVA, 2017.

Figure 04 shows, through the Treaty of Ayacucho, that the region at this time belonged to Bolivia and would therefore generate a lot of future friction as it began to be occupied by Brazilians in lands that were rich in latex, a raw material coveted at the time by the burgeoning car industry.

In 1877, the first wave of northeastern migrants arrived in the area. That year there were around 4,600 men, in 1878 more than 15,000 and by 1890 there were more than 158,000 men (MORAIS, 20008). When they arrived, these migrants encountered no obstacles when they crossed the border, as there was no demarcation of any land (TOCANTINS, 2001a). They threw themselves into the forest without the need to secure property rights because what mattered were the rubber trees and not the land. They lightly examined the riparian vegetation looking for rubber trees and defining the

along the region's main rivers and trade between the two countries was expanded (MELLO, 1990).

properties. Those who came later, when they found the mark of human presence, moved on in search of sites that had not yet been demarcated (CASTELO BRANCO, 1961; REIS, 1953).

With regard to demarcations, Oliveira (1985) reveals that they were not always harmonious, respectful and simple. In fact, they were the result of constant debates, various agreements and even the preparation of a war. The presence of the Brazilians made the Bolivians uncomfortable. And when the first Bolivian proclamations of possession of the area actually appeared in 1899, the Brazilians had already been there for more than 15 years, with large and productive rubber plantations and trading their rubber with aviation houses in Manaus and Belém and, through them, with the world's consumer centres. The Brazilian settlement of the upper Purus and Juruá rivers was already a fait accompli.

The fact is that when the treaty was signed, the geography of this enormous space between the Madeira and Javarí rivers was unknown. This resulted in various frictions between the border demarcation commissions, so that in the first identification of the Javarí River, in 1874, the geographical coordinates were: 07° 0T 17" south latitude and 74° 08' 27" west longitude, which proved that the headwaters of the river were not on the parallel of 10° 22' latitude, so the Brazil-Bolivia border had to follow the "straight" line (figure 05). For this reason, when in 1895, in a new protocol signed by the Border Demarcation, the coordinates identified in 1874 were considered to be the sources of the Javarí. This implied considering a vast area already occupied by Brazilians, as noted by Colonel Gregório Thaumaturgo de Azevedo, the Brazilian commissioner at the time (TOCANTINS, 2001a).

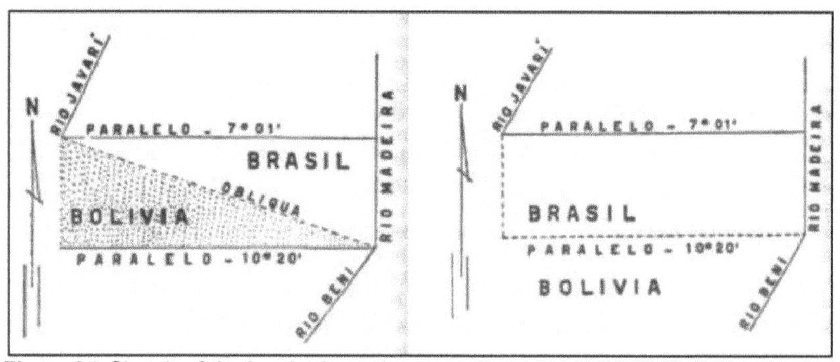

Figure 05: Sketch of the border issue between Brazil and Bolivia at the beginning of the 20th century

Source: Tocantins, (2001a).

Figure 05 highlights the change in the dividing lines between Brazil and Bolivia, which added to the space already demarcated in 1867. This would lead to complications in diplomatic relations, which were already strained by the settlement of Brazilians in their lands. Knowing the economic

48

importance of the Bolivian territories occupied by the Brazilians, and hitherto unknown to their people, as an offensive, in the 1890s, the Bolivians founded some rubber plantations in the upper river valley and were part of the same commercial network that transported all production via the ports of Manaus and Belém. This period also saw the arrival of Peruvians who settled in the Porto Carlos and Paraguaçu rubber plantations, now the municipalities of Brasileia and Assis Brasil, in the south-east of the state of Acre, carrying out extensive trade on the Madeira River, and were responsible for the development of the entire Beni, Orton and Madre de Dios valleys in competition with Brazilian rubber tappers. They were responsible for the entire development of the Alto Madre de Dios and Alto Ucaiale (CALIXTO, 2003; TOCANTINS, 2001a).

In the Juruá valley, in the far west of Acre, Peruvians began to invade in 1896 in order to set up some trading posts and extract rubber, and they approached the Brazilians themselves, who until then had not had any kind of disagreement.

With the caucho economy booming, more and more Brazilian migrants were coming to work in the rubber plantations in this region, which belonged to Bolivia under the Treaty of Ayacucho. Faced with this situation, Bolivia immediately set up a customs post in order to halt the Brazilian conquest. According to Morais (2008), the Bolivian government tried to secure what was rightfully theirs by taking possession of the lands south of the Madeira-Javarí boundary line in the early days of 1899, setting up a customs post on the banks of the River Acre, in the Caquetá rubber plantation, owned by a Brazilian and with the Brazilian government in agreement with the Bolivian initiative.

However, when he arrived in the Rio Acre region, a representative of the Bolivian government realised that the situation was much more complicated, as there was already a latex industry of various nationalities in possession of their properties with definitive or provisional titles. For this reason, it was necessary to guarantee the regulation of these lands. On 5 January 1899, Bolivian Consul José Paravicini, by decree, took some measures regarding the granting of land that were much to the displeasure of the Brazilian rubber tappers. In one of them, rubber tappers holding land titles were obliged to register them with the Bolivian Delegation Secretariat within a year and could no longer exploit them before obtaining the concession, otherwise they would lose their rights. Another measure also provided for the suspension of rubber tree felling from 1 August to 1 September, on the grounds that the tree needed to recover, and also recommended that when the rubber tree was all cut down, the rubber tapper would have to suspend felling for five consecutive years (TOCANTINS, 2001a).

This decree went against the grain of the rubber economy. In fact, it was a way of putting the brakes on the accelerated process that was developing intensely in Bolivian territory. This decision would generate an economic loss for the rubber tappers. And of course, the reaction was immediate, in resistance to Bolivian administration of the lands where Brazilian rubber plantations were located. According to Reis (1953), it would have been almost impossible to demarcate the rubber plantations within the established deadlines, since the process took so long and there were no surveyors.

From then on, as resistance, a series of events took place in the region. The first of these was the deposition of the then Bolivian consul, four months later, by a group of rubber tappers led by José de Carvalho from Amazonas. The act is known as the first Brazilian insurrection against the Bolivians. In a manifesto dated 1º May 1899, signed by fifty people, most of them rubber tappers, they demanded the departure of the Bolivian consul, which filled the population with hope and enthusiasm (MORAIS, 2008).

In other words, even though they were the rightful owners of this region, the presence of the Bolivians represented a threat to the strong rubber economy. As Morais (2008) reports, the Bolivians were seen by the Brazilians as invaders.

The second insurrection took place in the same year, when "the Independent State of Acre" was proclaimed by the Spaniard Luiz Galvez Rodrigues Arias. However, the action failed due to disagreements between the governments of Amazonas and Pará, the lack of support from the Brazilian government and opposition from Bolivia. He was deposed and the territory returned to Bolivia (TOCANTINS, 2001a).

The Brazilians' dream of securing this space didn't stop and resulted in the third insurrection, known as "the expedition of the poets". But military inexperience and a lack of organisation meant that the expedition was quickly defeated by Bolivian forces. The fourth insurrection, the armed and bloody phase of the Acre Revolution, began on 6 August 1902 until 24 January 1903, led by Plácido de Castro, culminating in the signing of the Treaty of Petrópolis that same year (figure 06) (TOCANTINS, 2001a). The Treaty of Petrópolis was signed in November 1903, bringing an end to the conflicts between Brazil and Bolivia. In the Treaty, Brazil undertook: to pay Bolivia two million pounds sterling; to build a railway between Santo Antonio do Rio Madeira and Vila Bela, at the confluence of the Beni and Mamoré rivers, as well as ceding a small area of land on the Abunã River and on the border with Mato Grasso (MELLO, 1990).

Figure 06: Brazilian commission for the Treaty of Petrópolis. Euclides da Cunha (last standing on the left), the historian, geographer and politician Barão Homem de Melo (with a white beard in the centre), as well as Rio Branco himself (seated, second from right).
Source: MELO, 2013.

Figure 06 shows the main authorities when the Treaty was signed in 1903, a fact that would end the issue between Brazil and Bolivia once and for all. However, the problem was not yet fully resolved, as there was still a border issue between Brazil and Peru, which also claimed land through the Treaty, as shown in figure 07.

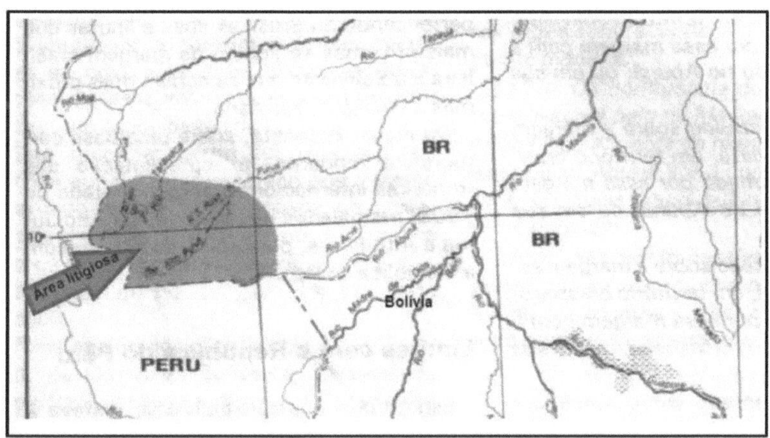

Figure 07: Dispute area with Peru
Source: MORAIS, (2008)

Like Bolivia, the Peruvian government tried to take possession of land in the Juruá and Purus

51

Valleys. Unlike Bolivia, there was no declared war, but two conflicts were recorded between 1902 and 1904: one in Funil and the other on the Amônea River. The border with Peru was only defined after the submission of reports by the joint *reconnaissance* commissions, set up to certify the *"uti possidetis"*[5] along the Purus and Juruá rivers. Finally, in 1909, the conflicts came to an end with the signing of the Treaty of Rio de Janeiro (figure 08), sealing Acre's international borders (MORAIS, 2008).

Figure 08: Map of Euclides da Cunha in 1909
Source: MELLO, 1990

Figure 08 highlights the map after the agreements signed between Brazil, Bolivia and Peru. The agreements of 1903 and 1909 brought about a definitive demarcation between the three countries, and Brazil returned a strip of land on the border between Acre and Rondônia, would also build the "Madeira Mamoré" railway - which was never completed - and would also pay a stipulated amount in kind. This would put an end to the territorial disputes in the region once and for all.

The various disputes over this territory, at that time rich in native rubber trees, from which latex was extracted, a raw material of great value that was emerging at the time. The region that would become the state of Acre is a place of intense trinational dispute. The interests of the three nations are clear: Brazil, Bolivia and Peru, the economic power that possessed the region. The issue was

---

[5] According to Reis (1948), *Utis possidetis* refers to the guarantee of lands already populated by any of the interested parties, preserving what had been occupied first.

purely economic between these nations.

The Brazilian rubber economy remained predominant on the international market until the year 1012, when Asian competitors emerged as the new leading exporters of this raw material (OLIVEIRA, 1985). At that moment, countless rubber plantations collapsed, an unprecedented crisis that had never been predicted before. This is the subject of the next topic.

3.3 Second migration from the northeast and the "Rubber Soldiers"

After a long period of decadence, two decades of depression and rubber inertia, which led many rubber plantations to go bankrupt, putting rubber tappers in total abandonment, suddenly a tragic variant: the Second World War. This was to reverse the situation in Brazil, and more precisely in its rubber plantations. The 1940s, although short, would be symbolic in terms of regaining supremacy in rubber exports, as it would once again stand out as a supplier of raw materials to the US, resurrecting and reactivating most of the rubber plantations. It was the second rubber boom in the Amazon and once again the intensification of migration to the region.

When the Second World War broke out, the nations of the world were in opposing military alliances: on one side were the allies (the former Soviet Union, the United States, France and others), and on the other (Germany, Italy and Japan). In the case of Brazil, which was under the Vargas dictatorship, the government opted for neutrality for a while until the well-known agreements were signed. The fact is that the occupation of Malaysia by the Japanese, allies of the Germans, had left the USA in a critical situation, since they depended on the rubber that Malaysia produced. (MARTINELLO, 2004)

The world conflict, which began in 1939, stretched from the Atlantic to the Pacific, passing through the Indian Ocean and the China Seas. Asia, rubber's main export competitor, had its plantations devastated by the war. The loss of the Malaysian rubber plantations, and of rubber, the raw material used and needed during the war, was an unprecedented defeat, as the Allied countries were unable to obtain supplies from the East. The only alternative now was the Amazonian Mediterranean. Thus, from one moment to the next, the Amazon found itself involved in the world conflict, due to the "forced" return of the wild rubber monopoly (BENCHIMOL, 2010). Malaysia had almost 100 per cent of its rubber plantations wiped out by Japanese bombing raids (MARTINELLO, 2004).

According to Silva (2010, p.82), "the Americans considered rubber, steel and oil to be indispensable products for maintaining the war industry, but after 1939, as the world conflict progressed, rubber received more attention from the government".

In this sense, it was once again necessary to relocate in search of this raw material and with it a vast contingent of labour. Brazil's involvement in the Second World War, in 1942, took the form of supplying military contingents for the fronts and also signed an agreement with the US agency responsible for controlling rubber production and exports, the *Rubber Reserve Company,* also signing the so-called *"Washington Accords[6] "* (BENCHIMOL, 2010).

These agreements included the following commitments:

1. Fixed price of US$ 0.39 per pound, put on board in Belém do Pará for a period of two years, with the right to readjustment during the following three years;
2. All rubber exported in excess of the annual limit of 5,000 tonnes would receive a bonus of US$ 0.025 per pound up to 10,000 tonnes, and from there upwards another five "cents" per pound, with the sums from these bonuses to be spent on health, research and development;
3. A contribution of 10 million dollars from the US government, of which 5 million dollars was for sanitation, to be carried out with the assistance of the Rockefeller Foundation, and 5 million dollars was to be invested by the newly founded Agronomic Institute of the North in scientific research, improvement and promotion of production. (BENCHIMOL, 2010, p.277)

The new situation was thus set: on the one hand, the urgent and short-term interests of the Americans, and on the other, the permanent and lasting interests of the Brazilian government, which wanted to set up a development policy in the Amazon to conquer the land, dominate its waters and subjugate the forest. According to Benchimol (2010), this battle involved armed strategies in the top ministerial echelons of the two governments, which established their policy and action by setting up a logistical-institutional apparatus of great magnitude for the time.

Once the agreements had been signed, with guarantees of the necessary basic assistance for the rubber soldiers to come, Brazilian government incentives began again, with intense advertising, optimistic posters and eye-catching slogans. To this end, it was necessary to hire specialists to help persuade these people to migrate to the Amazon, including the Swiss painter Pierre Chabloz. According to Moraes (2010), the artist, art critic, publicist, musician and cultural agitator, who arrived in Fortaleza in 1943 in charge of advertising for the "Rubber Campaign", played a significant role in the cultural environment of the state of Ceará, developing a deep emotional bond with the state. This was already part of the Vargas government's efforts to fulfil the agreements made with the US government the previous year.

Born in 1910 in Lausanne, Switzerland, Chabloz studied at the Geneva School of Fine Arts, where he studied Figure, Perspective, Graphic Arts and Decoration. He also attended the Florence and Milan Academies of Fine Arts and the Italian Brera Academy. He arrived in Brazil in May 1940, when he received an invitation that would allow him to participate effectively in the so-called "March to the West", working in the propaganda division of SEMTA (Special Service for the Mobilisation of Workers for Amazonia), a government body created in November 1942 as part of the CME (Economic Mobilisation Commission). In January 1943, Chabloz moved with his family to Fortaleza,

---

[6] The "Washington Accords" allowed the region to set up a logistical-institutional scheme, in which the Brazilian government actively participated, with American support, opening up many operational and strategic fronts in the area (BENCHIMOL, 2010).

the city to which SEMTA's recruitment centre was being transferred. It was in this context that Jean-Pierre Chabloz worked in SEMTA's propaganda division. In the six months dedicated to this activity - January to July 1943 - the artist developed a vast amount of material, including posters, booklets, illustrations for conferences, drawings of Northeastern biotypes, rubber trees and the rubber production process, as well as many preliminary studies. The poster certainly emphasises the abundance of Amazonian vegetation and the fertility of the land, which was supposed to function as a formal and semantic opposition to the aridity and scarcity of the backlands. In addition, the proliferation of rubber trees and workers conveys the idea of producing a large volume of rubber, which was in line with the government's objectives of transporting fifty thousand workers to the Amazon in five months through SEMTA, as well as raising premiums, granted by the United States, for every five thousand tonnes of rubber produced. (MAIA, 2014)It was the work of an intellectual elite forcefully promoting the idea that "the green paradise" was without a doubt the only solution to end the state of misery that afflicted the sertanejo people of the Brazilian Northeast. Although it only ran for six months, its production was so intense that it was historically marked.

Figures 09 to 12: Propaganda by the Getúlio Vargas government's Propaganda and Press Department (DIP), made by the Swiss painter Pierre Chabloz, which helped persuade northeasterners to come to the Amazon.

Source: SANTANA, M., 2012

Figures 9 to 12 show the advertisements designed and produced by the Brazilian government in the 1940s, during the Second World War, presenting the latex-rich Amazon region as a place that would produce great wealth. These were advertising methods of persuasive discourse, the responsibility of the Getúlio Vargas government's Propaganda and Press Department (DIP), to bring northeasterners back to the Amazon (SANTANA, 2012), now as "rubber soldiers". These images show the great power of the ideological image, presenting the region as a land of plenty, a land of victory, showing trucks carrying tonnes of rubber harvested by the workers, with no connection to the reality that awaited these workers. They were posters designed to convince the north-eastern region of Brazil. Perfectly articulated words, projected as an ideological sign, highlighting what Mikhail Baktin discusses in *"Marxism and Philosophy of Language"* (1999), when the word is designated by an ideological sign, and has the power to convince and instil persuasion in the individual.

The huge mobilisation did not spare even the then President of the Republic Getúlio Vargas, who published the following statement:

> Brazilians, with the same clarity with which I have become accustomed to speaking to you, I come to you today to ask for your loyal and determined cooperation in favour of a campaign that is being inaugurated today: the rubber campaign. You know how enormous the wear and tear is in the present war. (....) The Allied arms need more rubber, the rubber that exists(....). Extract rubber wherever you can, according to the plans that are being launched today in all Brazilian municipalities, with the sincere collaboration of your mayors. The solidarity of your

It was also common to hear daily appeals on the radio and in newspapers, as the newspaper

O Acre had already shown in 1943, according to Nascimento Silva (2000):

It was a discourse that, as Santana (2012) points out, needed to enter every home, persuade,
deceive and win over entire families to migrate to a "new world" that offered "comfort and dignity".
Official propaganda invaded the Northeast and the dream of "wealth and a prosperous life", of
"paradise lost", seemed closer than anyone could have imagined. The problems arising from life in
the caatinga would be overcome by a magic touch. With these speeches, the start was made for the
great "march to the West".

Northeasterners had to be brave and loyal, enlist and show their love for their country. These
extremely seductive posters had the important mission of fulfilling and strengthening the
determinations of the Estado Novo, through the Department of Press and Propaganda. They
became a part of everyday life: the man from the caatinga had to be convinced that he would become
the newest national hero and that he would certainly be living in an environment that would offer him
plenty, comfort and prosperity (MARTINELLO, 2004).

Ironically, parallel to all of this, again in the same way as the First Battle in 1877, the Northeast
was experiencing another major drought in 1942. According to Santana (2012), this factor
contributed greatly to the recruitment of these migrants, helping them to migrate to the Amazon.
Martinello (2004) estimates that the number of rubber tappers reached 34,000, with an average
annual production of 16,000 tonnes of rubber, but the Americans wanted to increase this production
to 45,000 tonnes in 1942, 60,000 in 1943 and 100,000 in 1944, which is why the government created
the "rubber battle".

In order to make this project viable, a number of bodies and institutions had to be set up
which, according to Nascimento Silva (2000), would be in charge of funding, recruitment, transport,
accommodation, medical and health care and food for those fighting in this battle. Among these were
the main ones:

Special Service for the Mobilisation of Workers for the Amazon - SENTA - later replaced by the Administrative Commission for the Forwarding of Workers to the Amazon - CAETA: its aim was to recruit, forward and place workers in the rubber plantations, transporting them to Belém;

Serviço de Navegação e Administração do Porto do Pará - SNAPP: In charge of transporting rubber workers from the ports of Belém to Manaus, Porto Velho and Acre;

National Immigration Department - DNIT: Its purpose was to recruit and send workers to the Amazon and also to supervise other bodies involved in mobilisation;

Special Public Health Service - SESP: This was responsible for providing medical and health care to rubber soldiers;

Commission for the Control of the Washington Agreements - CCAW: Created by Decree Law No. 4.523 of 25 July 1942, its function was to coordinate and assist the activities of the Brazilian and US groups that would work on the operationalisation of the rubber battle;

Banco de Crédito da Borracha - BCB (Rubber Credit Bank): Created by Decree Law No. 4,841 of 17 October 1942, it was in charge of carrying out credit operations, promoting production, financing extractive companies, as well as exercising the final monopoly on the purchase and sale of rubber, both internally and externally;

Superintendence for the Supply of the Amazon Valley - SAVA: Created by Decree Law No. 5.044 of 4 December 1942, its purpose was to supply the Amazon Valley with foodstuffs, coordinate the

measures to be taken with the states of the region with a view to supplying and increasing food production, providing for the acquisition of goods, both inside and outside the country, and their transport to the Amazon, forming stocks. (NASCIMENTO SILVA, 2000, p. 51-52)

However, despite all the creation of this institutional apparatus, and as is already well known, the people at the head of these bodies and institutions did not always carry out their activities satisfactorily, and many in fact contributed to the failure of the rubber battle. (MARTINELLO, 2004).In 1942, due to the drought, there were around 20,000 to 30,000 people living in Fortaleza, as abundant labour for the rubber plantations. At the end of that year and the beginning of 1943, the DNIT and the RDC managed to send around 15,000 people to the Amazon, made up of sertanejos from Ceará, Paraíba and Rio Grande do Norte. It was a family migration that "voluntarily" set out to cut rubber (NASCIMENTO SILVA, 2000). Many families were given only two options: either their children left for the rubber plantations as rubber soldiers or they had to go to the front to fight the Italians and Germans. Many preferred the Amazon (figures 13 to 15) (MARTINELLO, 2004).

Figures 13 to 15: recruitment of northeasterners during the Second World War

Source: MACHADO, 2005, adapted by SILVA, M.L. S" 2017

In figure 13, young men from various villages in the Northeast, in the capital Ceará, waiting to travel; figure 14, conscripted soldiers being transported in lorries; in figure 15, physical activity in the military regime to adapt the "inferiority" of the Northeasterners to the work of cutting rubber. As these images show, they were a veritable army of extractors, as Nascimento Silva (2000) puts it, when he says that they were like soldiers going to the battlefield in defence of their homeland. Thus, until they arrived at the Amazonian rubber plantations, these "soldiers" had to travel a long and arduous road, either by land or by water, from the north-east of Brazil to the most distant rubber plantations, as was the case with those who came to Acre, as map 02 highlights.

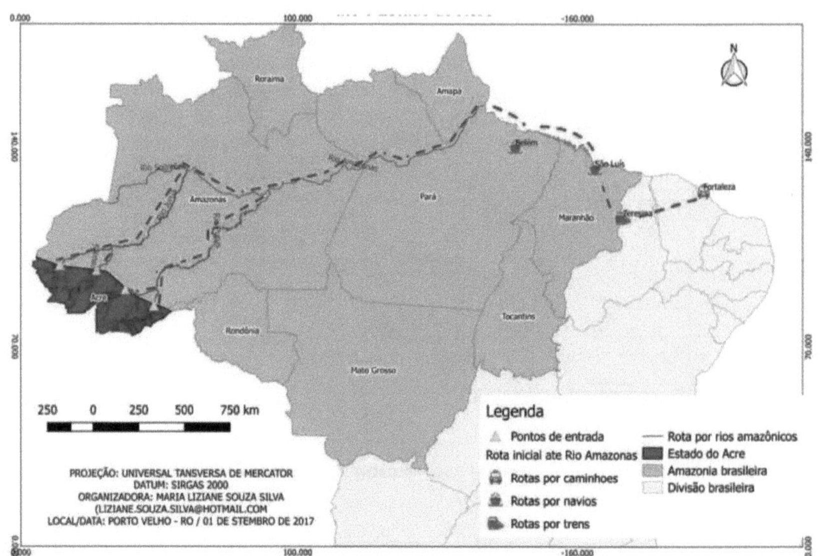

Map 02: Migration route of northeasterners to the state of Acre during the Second World War

Map: Maria Liziane S. Silva, 2017.

Map 02 illustrates an idea of the route travelled by northeasterners until they reached the port of Belém. This was only the first part of the journey, because after Belém there was still an arduous road to travel until they reached Manaus, then Acre and finally their rubber plantations. Benchimol (2010), an author who saw and lived through this moment, tells us about the arrival of these migrants at the Ports of Manaus:

> It was 1942. °While attending the 2nd year of the Faculty of Law of Amazonas and working as a baggage handler on the Panair do Brasil float, on the Manaus Harbour Roadway, I watched the arrival of numerous waves of immigrants from the Northeast who arrived there daily, attracted by the publicity hype of the famous Battle of the Rubber as a result of the Washington Agreements, signed shortly after the Japanese invasion of the Malaysian rubber plantations. (BENCHIMOL, 2010, p.203)

It's no coincidence that Benchimol describes an Amazon that was, in fact, formed in the image of the people of Ceará and the rubber tree, the keys to its social and economic formation.

However, at this point in the migration, something different was happening from the First Battle: now, as well as people from the north-east, people were also migrating from other regions of the country, including from public prisons. In all, 11,000 men were sent, but only 900 were placed in rubber plantations and the rest stayed in Belém and Manaus. These men were always causing trouble and messes, giving a bad name to the image of Northeasterners, and from then on, all the rubber soldiers started to cause fear in city dwellers. (NASCIMENTO SILVA, 2000). In Martinello's (2004) contributions about these people, the author emphasises that they were people of all classes, colours, professions and ages, strangers to their own environment and economic history. According to him, in Manaus, the Bulletin of the Commercial Association of Amazonas, of February 1949, describes the harsh and dramatic quality of the elements that came to the Amazon:

> What the Amazon saw at that time, constituting the majority of the rubber soldiers who arrived here, was not the fine flower of the backlands, nothing like that, but the mud of the asphalt, the riffraff of the big cities, even those who had just come from prisons, and who, arriving here, robbing and killing, looting and injuring, using all the processes of violence, began a stage of terror and crime, seriously damaging the peaceful habits of the population. What the Amazon saw, with a few small and honourable exceptions, was the malandro from the slums of Rio de Janeiro, the sangrador from the caatingas, the murderer who had done time in Fernando de Noronha or Ilha das Flores, in short, the Brazilian social scum on the streets (MARTINELLO 2004, p.246).

As you can see, the rulers were not concerned about the safety of those who had honourably come to "do their duty for the Motherland" because they had dumped here the social scum of Brazil, the "mud of the big cities". In addition, at that time, criticism rained down on the management of the organisations created to support these migrants. Among these criticisms was one by the former interventor of Acre and former president of the Rubber Bank, published in a newspaper in Bahia in 1943, in which he made a critical analysis of the chaos generated by the disarticulation and distortion of objectives, according to Benchimol (2010):

> The Battle of the Rubber is a problem of organisation. Organisation of rubber plantations, supply lines, transport, assistance and, finally, credit. What caused the failure? The answer is only one: because the interests of certain individuals and the incompetence of others maintain the prevailing disorganisation and lead everyone to disbelieve, because the "fifth column" takes advantage of this state of affairs and "directs" rubber tappers and rubber tappers towards its goals.
>
> The thousand and one organisations speaking in different languages form a veritable Tower of Babel. Semta, now replaced by Caeta, Sava, Sesp, Snapp and Bancreva don't understand each other. Semta brings the northeasterners to Belém and hands them over to Sava to feed them and send them to the rubber plantations. Semta doesn't know if Sava has food and accommodation available, nor do they know if Snapp can transport the men, luggage and materials upriver in good time. Neither of them takes into account the proper time to cut the rubber, nor the feasibility of transport, which must be carried out at the right time for each region, dictated by the periodic floods and terrible ebbs and flows of the rivers.
>
> [....] It is common for supplies to be unavailable for transport upriver when navigation is possible, and for goods to rot in Belém and Manaus because they have arrived there when the upper rivers are dry. The result: a year of deprivation and no production.
>
> [....] Men are supplied like "things", to whoever asks first, without knowing if the rubber plantations are ready to receive them, if the tools have arrived and if there is a way to feed them.
>
> [....] Discouragement sets in quickly. Defeated, they return to the cities, often on foot, to join the legion of the unemployed and beggars. Those who stay resign themselves to the fate of always being in debt....They receive four or five cruzeiros a kilo for rubber, when it is priced at 25 cruzeiros. [....] Hunger, sadness and disillusionment. (BENCHIMOL, 2010, p. 280-281)

It can be seen that the many promises made by the countless and convincing adverts have been transformed into beliefs, forming only part of these people's imaginations. The quote above demonstrates this situation, attesting to the fragility of the defeat, the ineffectiveness of the management of these bodies, total chaos, in the face of the sacrifice of thousands of northeasterners convinced by the appeal of the Brazilian government.

It's true that many of those who opted for the Amazon believed that they would be taking fewer risks in life and were also counting on the possibility of getting rich by producing rubber and then returning home victorious. When they saw the misleading adverts, they were already inside the rubber plantations. It was said both in the newspapers and in the paintings on the walls of the northeastern cities that the rubber tree bore a lot of fruit, the size of a ball, and the rubber tapper's only job would be to harvest this fruit, which cost a lot of money, making it easy to get rich. All these attractive lures contributed to thousands of migrants travelling in dangerous conditions, because as well as being long, the journey was tiring. On overcrowded ships, without the slightest comfort, there were thousands of people, including men, women and children, creating chaos and turmoil. The food was of very poor quality. Many arrived sick, others fell ill in the inns and were thrown there, crammed "like animals", suffering hunger and humiliation. In this way, more soldiers died than in the army of the Brazilian Expeditionary Force fighting in the fields of Italy in 1944 and 1945. Thousands of rubber soldiers were exterminated by the diseases that debilitated them without having the minimum assistance promised in the campaigns through the agencies created, victims of the neglect of the government and its representatives (NASCIMENTO SILVA, 2000).

With the end of the Second World War in 1945 - after Germany and Japan surrendered and the latter was hit by two atomic bombs, one in the city of Hiroshima and the other in Nagasaki - came the end of the Second Amazon Rubber Cycle. Asia, Brazil's main competitor in latex exports in the following years, gradually regained its production, which had been devastated by the war, and consequently its position as the leading exporter of gum. In this situation, the fall in Brazilian rubber exports was nothing new, leading to the decline and bankruptcy of many rubber plantations once again. (MARTINELLO, 2004).

This is how the northeastern population was formed in the Amazon, which, in addition to the reasons we have already seen, i.e. escape from the scourge of the droughts, the argument most used by the enticers was that it was the only alternative to fleeing the call to fight in the camps in Italy. The *Washington Accords* between the Americans and the Brazilians, which covered the period from 1942 to 1947, highlighted different interests: while the former was marked by opportunism and pragmatism, the latter was for the implementation of a long-term development programme in the Amazon. This second rubber boom did little for the Amazon in economic terms, given the deplorable situation of the thousands of migrants there after the war. A great deal of propaganda was invested in material and symbolic resources to attract these people. Many took refuge in the centres of the big cities, forming a belt of misery. As an example of this, Martinello (2004) highlights the exodus from the rubber plantations to the city of Rio Branco, which was intense in the years that followed. With the end of the extractive economy, these migrants arrived in the city in search of better living conditions and job opportunities. They live mainly on the outskirts of the city, making up the contingent for the process of countless invasions in search of a place to live. The immediate problems of this migratory influx were the emergence of clandestine allotments in places without any infrastructure and the proliferation of informal labour. The city was therefore unable to absorb all these people, causing major problems (MORAES, 2008).

As at the end of the first rubber economy, these rubber tappers were left to fend for themselves. State support was ineffective because the system had no interest in "minor" causes and the only option for survival for many was the clandestine refuge of the city.

With this in mind, the next chapter will look at the narratives of some of these migrants, and/or their children and wives who cut rubber in Acre's rubber plantations, using the DSC methodology, where we analyse the collective statements.

CHAPTER 4

THE VOICE OF THE "INVISIBLE"

I saw an old rubber tapper lamenting

From the country you once knew

He lived his whole life on the rubber plantation

On this ground that years ago you came to live

He insists he doesn't want to believe it

In the forestry scenario you're seeing

The habitat it lived in is dying

Dishonouring your homeland and your country

The pain in the chest hurts and the scar bleeds

Listening to the echo of the rubber tree groaning.

(Francisco Marquelino Santana)

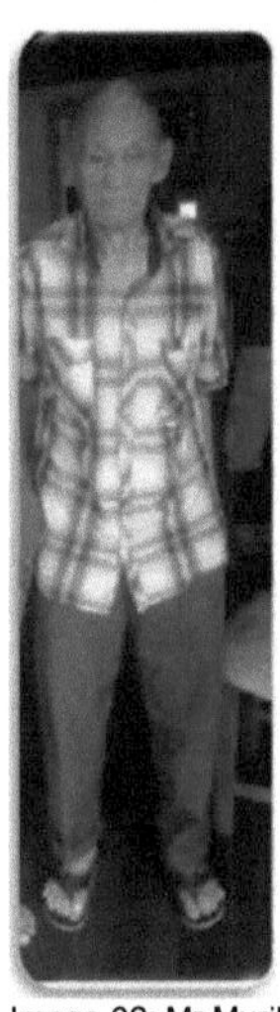

Image 02: Mr Murilo de Lima, *(Inmemoríam)*

Photo: SILVA, M. L. S, 2017.

This chapter will deal with the Discourses of the Collective Subject - DSC, where we will present the discourses obtained through the selected "central ideas", duly represented by categories, found in the answers and comments resulting from the interviews. The aim is to deal with the speeches of northeastern migrants through this technique, developing their opinions qualitatively by tabulating and organising this set of speeches into a collective thought. The idea is not just to tabulate, but to give existence to these memories (BARBOSA, 1994, p. 12).

Subsequently, the results of these interviews will be discussed and analysed in conjunction with the ideas of authors who work on and discuss this issue.

4.1 The Discourse of Collective Subjects: the speech of people from Ceará and their descendants

During the interviews, ten questions were put to each of them, which resulted in a variety of answers, sometimes divergent from one another, but mostly in agreement, represented in the following topics.

Question 01: Why *your family came to Acre*

The aim of this question was to find out what motivated the person or family to leave their area, their region, and migrate to another so far away and with different characteristics.

In this first question, six categories were found in the interviewees' speeches (graph 01), with category B standing out - they came during the first rubber economy - deducing that the majority of these interviewees are people already born in this region, but descendants of northeasterners who arrived during the first rubber economy at the turn of the 19th to the 20th century.

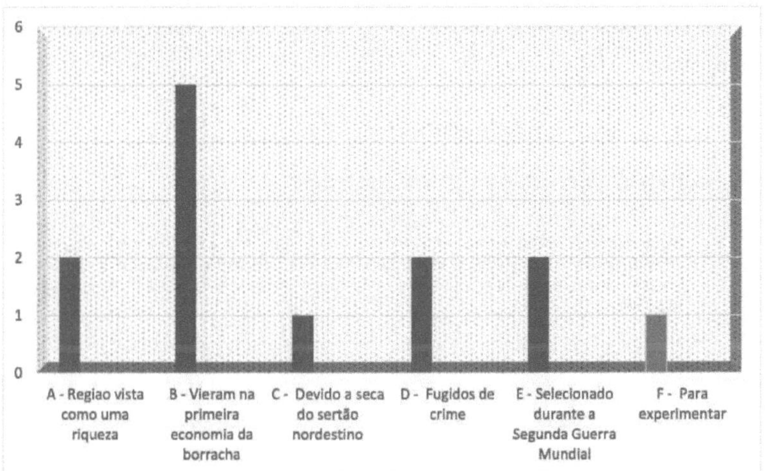

Graph 01: Distribution of categories for the question: Why your family came to Acre

Source: Prepared by SILVA, M. L. S based on interviews conducted in February 2017.

Graph 01 highlights category B, showing that most of the interviewees are from families that came during the first wave of migration to the Amazon/Acre and are therefore born in this area.

CATEGORY 01 - A

DSC1 A: That here was a river of wealth

> Yes, they used to say there (in the Northeast) that here was a river of wealth, rubber was wealth, everyone who was a rubber tapper was rich, had a lot of money!; My father used to say that they used to say that the rubber tree was a bag of money! Everyone lived well and the people from Ceará came here, a lot of people from Ceará at that time! Then they'd say that in the Northeast, in Acre, you collect money with two squeegees! The bosses here used to send for them and say: rapaz rumbora pro Acre cortar seringa, lá é bom rapaz, num falta nada, tem muita mercadoria (DSC elaborated with the testimonies sd1; sd2;)

In these speeches we can see the power of large government advertisements and/or the bosses themselves in search of labour for their rubber plantations. Northeasterners were the target of these adverts, of the ideologies created about this region.

CATEGORY 01 - B

DSC1 B: they came during the first wave of migration

> My father was one of the first here. They came in the first wave (the first wave of migration); my grandparents are all from the Northeast. My father, Joaquim, arrived here in Rio Iaco, in Sena Madureira, to cut 1906. But my grandfather, my mother's father, Jose Alves de Oliveira, was the one who explored the Itamaraty and Santa Clara rubber plantations on the Iaco; My grandfather was from this group from there, from this part of the world, from Paraíba, they came on the first cut; He talked a lot about here, there were my guys who went from here to there, he told me, one was Vintura, his name was Vintura Anastácio and the other was..., I don't know.., I've forgotten the name of the other... this other one was my legitimate cousin, he worked here in Envira, then he left two married daughters there and took two boys to Ceará (DSC made up of all the statements sd3; sd4; sd8; sd9; sd10)

Most of these interviewees are migrants from the first rubber boom, when the region was a pioneer in latex harvesting.

CATEGORY 01 - C

DSC1 C: That the sertão was dry, people died of thirst and hunger.

> My grandfather and grandmother were from Ceará. They told a lot! She told me a lot about when they came here by ship at that time. At that time, people from the sertão, it was drought, people were dying of thirst, of hunger, they came in the first wave. (DSC elaborated from statement sd4)

For many of them, going to the Amazon would be the chance of a great financial "turnaround" due to the drought problems experienced in the northeastern hinterland, and the Amazon would be the place to solve everything, "the paradise that had been found" and would thus solve these problems.

CATEGORY 01 D

CSD 1 D: Escaped from crime

> Well, they say that when these Afragelados came here from Ceará and never went back, it was because they were criminals. But then, when they leave there, they're already damaged, right? They're going to tell us about their lives? they can't, they can't. They'd even change their names. Because a lot of boys and girls came; a lot of runaways came, my father told him, my grandmother told him. A lot of runaways. My husband was a very tricky person back in Ceará, my daughter. When I married him, he had his documents sent to the registry office, and he had

66

changed his name here. So he kept his new name. (DSC based on statements sd1, sd5; sd8;)

As you can see, the Amazon wasn't just a refuge for people who were experiencing problems due to the climate in the north-east, it was also a place where many criminals who had fled their regions sought refuge in an attempt to start their lives over.

CATEGORY 01 E

CSD 1 E: Selected during the war: Choice of rubber tapping, because of the greater chance of survival than the war

> My husband, Eufrásio, came from Ceará, because of the war, he was one of those called up, he came at a time when a lot of people came (the Second World War). He was drafted, because either you went to war or you came to cut rubber, it was a matter of force, because there was no escape, he was the oldest in the family, he was the one who was drafted, if there were two, one would go; I came to serve the government there, cutting rubber, ne! I had to fulfil my obligation. I was a soldier! When I received my uniform, that's when I got the news that I'd been selected for the war. So they made the deal for me to serve here. I came because of the war. I was a beast! Everything was afraid of me. Here I'm a good piece of work. I was respected by everyone. Sometimes, I wonder how a person wins a battle like this, and arrives at this age, right? Then, when I was due to leave, I was eight days away from getting my uniform. Then they asked me what I would choose, whether to come here or serve there, go to war. Either come here or go there. I was mad. Then I said: boy, I'd rather go to the Amazon, because when I get there I'm guaranteed a life, and in the war I'm not guaranteed, (laughs). They transferred me, they sent me here (DSC based on statements sd6; sd9).

During the 1940s, the period of the Second World War, there were also those recruited for the battle with a choice between going straight to the front or serving as a "soldier" cutting rubber in the Amazon and, because they didn't have much choice, many preferred to serve in the Amazon rather than go to war because they thought they had a better chance of survival.

CATEGORY 01 F

CSD 1 F: To experiment

> My grandmother came straight to Amazonas. She died when I was little. Boy, I don't remember much. She said she had come here because it was very bad there, the drought, it was very strong, so she saw that it was better here in Amazonas, so they looked for her here in Amazonas. They came to try it out, then came back, but they didn't come back. (DSC elaborated from statement sd7)

Diverging from the objectives of the majority of migrants, there were still a few who opted to come to the Amazon as a "gamble", but who were seduced by the countless adverts, just to "try it out", as they intended to stay for a short time.

Thus, in response to question 01 "The *reasons why your family came to Acre"*, we noticed that the subjects selected here were almost entirely descendants of northeasterners. This means that they are children and grandchildren of this migratory process, no less important of course, since they were born into this world and are as much a part of it as their parents, since they have experienced every moment of their lives in this space. However, there are also those who are still alive, albeit at a more advanced age, to tell us their story, which makes this research more attractive and meaningful by being able to recount such a wealth of detail from their memories, which one gradually realises are being lost over the decades in more distant spaces.

These thousands come from the first and second rubber economies, which began at the end of the 19th century and were reactivated in the middle of the 20th century. Many were convinced by the countless advertisements for the "land of wealth", and also helped by the drought that ironically hit the region during these two periods. The Amazon thus appeared to be the solution to these problems, since the possibility of enrichment was taken for granted.

But these were not the only reasons for attraction, many of these migrants were fugitives from crimes committed and used the region for refuge and work. Others, however, came just to take advantage of the opportunity to find a good financial return, but soon to return home.

But as can be seen from their speeches, regardless of the reasons that drove them to migrate or the time, what they had in common was the hope of a better life, of being able to provide better conditions for their families.

Question 02: *What was the situation like inside the rubber plantations?*

The aim of this question was to understand what their situation was like when they arrived here, their adaptations, their experiences, the internal rules of the rubber plantations, and what their new life was like in the land of fortune.

On this question, five categories were obtained from their speeches (see graph 02), with category A - Very difficult - standing out, revealing that the situation for almost everyone in the rubber plantations was one of total sacrifice.

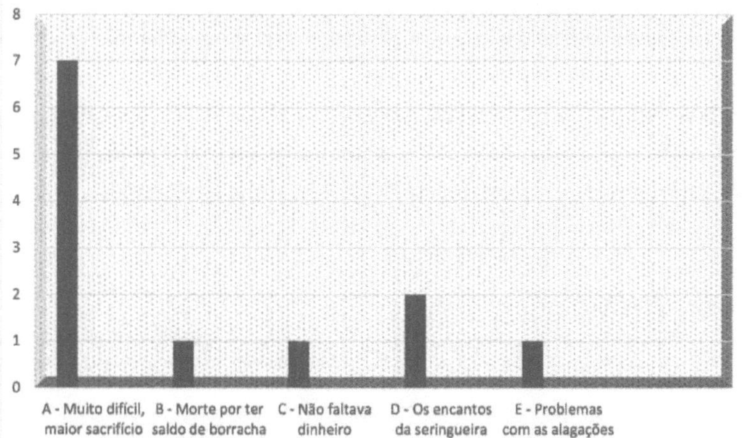

Graph 02: Distribution of categories in relation to the question: What was the situation like in the rubber plantations?

Source: Prepared by SILVA, M. L. S based on interviews conducted in February 2017.

The graphic highlights the total contradiction of the countless advertisements that existed in the northeast, which omitted the raw and wild reality of the region, which was always idealised as a good, easy land of abundance for those who ventured into it.

CATEGORY 02 - A

DSC2 A: Very difficult, the greatest sacrifice.

> The people from Ceará who came here just came here to suffer, to die! They just came here to die and suffer! There wasn't one person who said they lived well! suffering in the forest, cutting rubber, hungry, leaving home at midnight to cut rubber, like my father did,...everything here was rough forest, every corner, there was no road to anywhere, rubber tappers carried rubbers on their backs, sacks of flour on their backs, hours and hours to get home, to their little shack. It was a great sacrifice, I was poor at that time, it was such poverty that it was a pity. They didn't know..., they wanted to and they killed the roads, they cut them without knowing, the road died. The people were very poor, they were covered in patches, their clothes only had patches, from the mouth of their trousers to the top, at the waist, they were patched, it was men and women, there were few people who didn't have patches on their clothes...; It was more difficult in those days
>
> There was no vegetable garden, no fridge. And so..., everything was difficult, everything was very difficult; back then anyone who didn't work lived in misery, there was no one to give anything to! If you worked, you still lived with sacrifice! If you worked, you still lived with sacrifice! Working in the sapecado, in the danado, it was in the fields, it was in the rubber plantations, it was in the bush, it was in the shellfish, everything was with sacrifice to get that tiquim...., if you wanted to live!; Ah, in those days things were difficult, in those days it was.., things were very difficult in those days; It was difficult, it was difficult in those days,.... we'd leave at dawn and arrive at one o'clock, three o'clock, there'd be a long, long road, after arriving we'd go to get the chips, those thick sticks they'd cut and get the chips, or we'd go to the forest to get coconuts, butted them in the furnace, and then smoke them. They suffered..., they suffered a lot; it was difficult, difficult, it was. In the rubber plantations they made rubber, took it to the shed, then sold it to the boss, then bought the goods he wanted, and went home, and sometimes we made those little balls of cernambi to sell at the regatta; When the rubber plantations went bankrupt there were people who almost died, with a terrible crisis, they had

69

<blockquote>nothing, and the money was gone too. Brazil's money ran out, which was Brazil's gold. (DCS elaborated from statements sd1; sd2; sd3; sd4; sd6; sd7; sd8)</blockquote>

These speeches show that the grandiose propaganda fell flat the moment these migrants arrived and encountered the reality of the rubber plantations. There was no ease for anyone; on the contrary, hardship prevailed. There were hardships, dangers, hunger, it was a time of great suffering to work, to survive. At this point, the discourse of "the land of the bill, the land of money" fell to the ground, which may have been true for others, but not for those who had paid a high price.

CATEGORY 02 - B

DSC2 B: death due to rubber balance

<blockquote>The late Mâncio had a lot of people killed here, everything was killed so that there would be a balance. My grandparents ran away from the rubber plantations to avoid dying. If you had a balance, if you didn't pay the balance... well, that's nega véa. (CSD elaborated from sd5 testimonies)</blockquote>

Another strange and frightening situation that existed within the rubber plantations was the possibility of death in the face of the balance of the rubber (still in the first rubber economy, at the beginning of the 19th century), in the face of accountability to the boss, revealing the rigid systems operated by the "colonels" who acted as if they were Laws, for everything and everyone. This discourse reveals the treatment of some rubber tappers towards their employees, which is contradictory to the promises made by many when they were looking for labour in the Northeast to work in their rubber plantations - a perverse, inhumane system of total financial and social control that existed in this area. Oppressed, these rubber tappers were slaves to their own fate.

CATEGORY 02 - C

DSC2 C: There was no shortage of money

<blockquote>The syringe was Brazil's treasure, Brazil's gold! It was the movement of Brazil. Even here the movement was different, it was busy, because rubber was sold, sometimes in the morning you didn't have any money, but in the afternoon you did, because everyone wanted to buy rubber. It was a crazy movement here, and there was money in our pockets for those who were rubber tappers and there was no shortage of money for merchandise, only those who were lazy were in need. (DSC elaborated from statement sd8)</blockquote>

Contradictory to most of the other speeches, there were those who were lucky enough not to be short of money. The speech *"The rubber tree was Brazil's treasure, it was Brazil's gold!"* reveals the idea of a good, bountiful time, which for this individual was worth the whole rubber boom. In addition, there is the difference between one era and the next, where in the early days (the first rubber boom) there was the possibility of the rubber tapper's death when he made a loss on the labour obtained from extracting latex, but in this second period (the second rubber boom, from the 1940s onwards), there was a clear change in the internal systems of the rubber plantations. Workers had more freedom in their earnings, showing an improvement in the treatment of bosses and rubber tappers. Many of them made good money through the "treasury of Brazil".

70

However, those who thought that cutting the syringe was a simple act were mistaken: it had to be "uncapped" before the latex could be extracted.

CATEGORY 02 - D

DSC2 D: The charms of the rubber tree

> Boy, even today there are people there who don't know what a syringe is. The syringe is enchanted, all syringes are enchanted. It's the woodsman who disenchants it. Wood that only gave a book of milk, I'd go there and let it spill.

> I knew how to do it. Look, when the lights came on, the first day, I'd scrape the road three times, now if there were three scratches, I'd cut in three scratches, then in four, I'd go there and scrape again, which is, three scratches she increases all the time. Leave it where you want it (laughs). I couldn't cope with twelve cans. It was a lot of weight and I only arrived at night.

> I was lucky with the roads. I only took good roads. I already knew about the experiences that taught me; where we worked, the work was difficult, you only woke up at dawn. But I never liked working at dawn. I tried cutting with a torch, with a torch, with a torch, but I didn't do well with any of them, so I only cut when it was early in the morning. I used to cut at dawn because it was better to get the milk out, it's colder, and the syringe has the wind to pick up the milk straight away. But people got less milk during the day, like me. Sometimes I went at dawn. There were roads that were far away, so I'd go. The annoying thing was leaving very early, very sleepy. We made the shoes ourselves. But then your feet would get red and shiny. The clothes were sewn by the woman, who bought fabric in the shed. They weren't washed every day because there wasn't time, sometimes they were only washed on Sunday. (DCS based on statements sd9; sd10)

The accounts of these individuals show a very peculiar situation: the need for the correct knowledge to cut the syringe, otherwise it would be time wasted, a waste that could jeopardise everything. The syringe "with its charms" shows the rubber tapper the need for malice, empirical knowledge about its treatment and handling. It had to be "disenchanted"! This reveals the knowledge contained in these individuals, acquired on a daily basis and over decades through their experiences in the rubber plantations. Care, perception, intuition and intimacy with the rubber tree were essential for a good latex harvest. Without this everyday knowledge, nothing could be gained.

CATEGORY 02 - E

DSC2 E: Problems with flooding

> Fia, in the summer, it was very good, you had everything, right, now when the flood was early, you ate the fields you had planted, sometimes there was verdiim, the flour houses were too little for too many people, many asked for it, when it was early they were out of flour. And many people ate green bananas, I was one of them. One year the floods came early, then all our fields rotted, but we still made a little, then the flour was gone. Then we ate it with green bananas, with pamonha, couscous and jerimum. The only thing I didn't eat was palm hearts, like a lot of people did, acai palm hearts. It came from the pupunheira tree. Back then it was for those who couldn't afford it. Today it's only for those who can, (laughs) When the manioc started to get thick, I would grate it and they would make that little kiss the size of a pirão, she would do that. When it wasn't, I'd cook the manioc, then put it in a pot like this, then Dad would pound it in a mortar and pestle and make a pirão for everyone. He'd roast the fish and we'd eat it, I thought it was good. (DSC elaborated from statement sd7).

The problem of flooding was and still is a fairly common situation for Amazonian populations, especially those who live near rivers and streams or need them to get around and survive. Periods of flooding in the Amazon can be very disruptive for many residents. The periodic floods can directly or indirectly affect these rubber tappers with large or small losses, either through the loss of their crops or the lack of access. The result was great food deprivation for those rubber tappers who lived off this type of subsistence.

Question 02, *"What was the situation like inside the rubber plantations?"*, reveals two situations inside the rubber plantations: the first, during the first rubber boom, at the beginning of the 20th century, where the Laws were governed by truculent, violent and arrogant bosses who took away the possibility of making a living by extracting latex. Prevented from having their own subsistence plantations due to the intense pace of work on the rubber roads, these workers were subjected to the strict system that prevailed in this area, making the situation within the rubber plantations even more difficult. Poverty prevailed at this time, hunger was persistent and resources were scarce.

In the second phase of the rubber economy, after the 1940s, this situation began to improve, as the phrase "there was no shortage of money" shows, and the relationship between boss and rubber tapper began to improve. Now, unlike before, this rubber tapper can have his own swidden and is no longer totally dependent on latex extraction alone, as was the case in the first economy.

But at all these times, nothing was easy, everything revolved around difficulty, whether due to climatic issues, when there were floods, or social issues within the rubber plantations. In addition to all this, it was also necessary for this rubber tapper to absorb a certain intimacy with the typical knowledge in order to be able to work with the rubber tree, otherwise he would be condemning the rubber tree, his source of livelihood, as he says: "they didn't know....and they killed the roads, they cut without knowing, the road died".

Question 03: *What was the food like in the rubber plantations?*

The idea here is to get their accounts of how they worked and got their own food, at the same time as the arduous working day within the rubber plantations, given that they lived far away from everything and everyone. 03 categories were found in this question, as shown in graph 03, where the category - From the forest and the river - stands out as the main sources of their food.

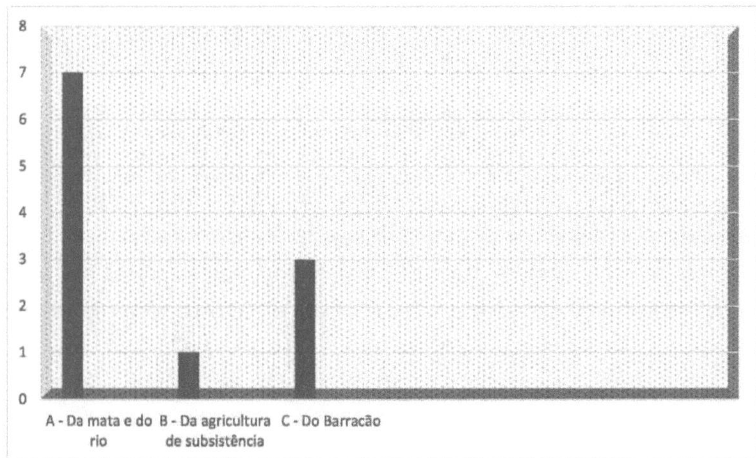

Graph 03: Distribution of categories in relation to the question: What was the diet like in the rubber plantations?

Source: Prepared by SILVA, M. L. S based on interviews conducted in February 2017.

Graph 03 shows a typical feature of life for residents of the dense Amazon forests, who use the forest and the river, through hunting and fishing, as a source of food, since it is not common to set up businesses in this area due to the distance and the great difficulty of transporting them.

CATEGORY 03 - A

DSC3 A: the forest and the river.

> It was from the forest, only the fish from the river and the forest, we didn't eat anything else, everything was from the rough forest; It was from the lakes, from the forest, fishing, hunting. Even on the road, we killed animals, it was difficult to go to the forest, it was on the road; What we lived on, the food was from the forest because they killed the embiara for us to eat. Those who had chickens were fine, those who didn't killed nambu, that was the food of the people there; Nobody bought food in Seringai, nobody did, they went to the road, they killed nanbú, cutiara, quatipuru, they killed everything. Big game too, deer, pigs, they killed everything, but nobody ever bought a ranch. We used to see jabuti, blue nambu, chicken nambu... so many embiara. Once we came across a flock of peccaries, I almost died of fear. My husband saw a trail of jaguars; so..., there were lots and lots of fish. There was also a lot of game, in those days there was a lot of game, it wasn't invaded, there was a lot of fish, a lot of meat, everything, tapir, there was a lot. At that time we lived off hunting. We killed a lot of game and lived off it. I thank God that when I cut rubber, all four of them were good, there was game in the forest and fish in the river. Everything was good. When the hunter doesn't kill on the road, we take a day off to hunt because we can't be without a ranch, but it was difficult, there were lots of deer, pigs. In the places where the peccaries lived, it was clean, clean. (DCS based on statements sd1; sd4; sd5; sd6; sd7; sd8; sd10).

This discourse shows that the forest and the river were the main suppliers of food. There was plenty of game and fish, without the need for much effort. Game, for example, was often caught during the cutting of the rubber tree, as there were many wild animals in these places.

DSC3 B: subsistence farming

> Poor, they had nothing to live on, they raised a bacuri on a rope, they raised a chicken, but they couldn't raise much because before it grew it would eat itself and they still had to cut rubber. Others lived on agriculture alone, planting a small plot of land, corn, these things, everything was very difficult, to buy things was a sacrifice, there were people who ate until they were sick, there was a lack of flour in everything. (DSC elaborated from statement sd1).

For many, subsistence farming was the guarantee of not running out of food (cultivated only from the second rubber economy onwards), but not everyone was as lucky due to the hard work involved in cutting rubber and the lack of cultivation. The result was a lot of deprivation and need for access to food.

DSC3 C: from the shed

> ...then the flour was bought at the depot..., except everything was in the shed; When I went self-employed, my own merchandise was in a bucket! A rubber tapper who was good would carry merchandise for the whole year. We bought everything outside. Everything was in the bucket; the rubber tappers always left at the beginning of winter, to get the year off. The boat would come to the harbour and pull in. There he would receive the goods to get through the summer, the flour, which was water flour that came from Pará at the time, then sugar, coffee, milk, jabá, tinned fish, pirarucu, there was always that in the shed. There was the comboieiro, who kept the animals, who took food to the rubber tappers, and brought the rubber, he was the middleman. Transport was by donkey. I was a comboieiro when I was thirteen. But it was like this: when the rubber tapper arrived at the rubber plantation, the comboieiro would take the food, he would take it twice, and the third time, he would only take it if the tapper had rubber. If he hadn't produced, he'd come back, even if he needed it. The rubber tapper's placement was paid for by the rubber tapper. When he wanted to leave, the boss would pay for the improvements he had made. He bought it from him to sell to another rubber tapper. If he hadn't done anything in three years, the boss would send him away, the train driver would take the order, the scales. If the rubber was enough to pay for the goods, he'd stay, if it wasn't, he'd leave. (DSC elaborated from statements sdõ; sd8; sd10)

The Barracão was also the place where these rubber tappers were supplied with food by the rubber plantation owners, so as not to leave their employees destitute. However, there were rules whereby, if there wasn't enough rubber, the comboieiros, the people responsible for delivering the products, were instructed by the rubber tappers not to allow the goods to be delivered and, once in debt, the rubber tapper would be "invited" to leave. But those rubber tappers who were considered "good" would take the merchandise for the whole year and still secure their allotment. The question *"What was the food like in the rubber plantations?"* shows that despite having the help of the sheds, the forest and the river were the main suppliers of food for the rubber tappers, and subsistence farming helped them make ends meet. The abundance of hunting and fishing at the time made it easy to obtain food.

Question 04: *Danger of attack on rubber plantations*

The idea is to find out from these accounts the dangers faced on a daily basis by rubber tappers who ventured into the immensity of the forest in search of rubber. 03 categories were found in the interviewees' speeches, as shown in Graph 04.

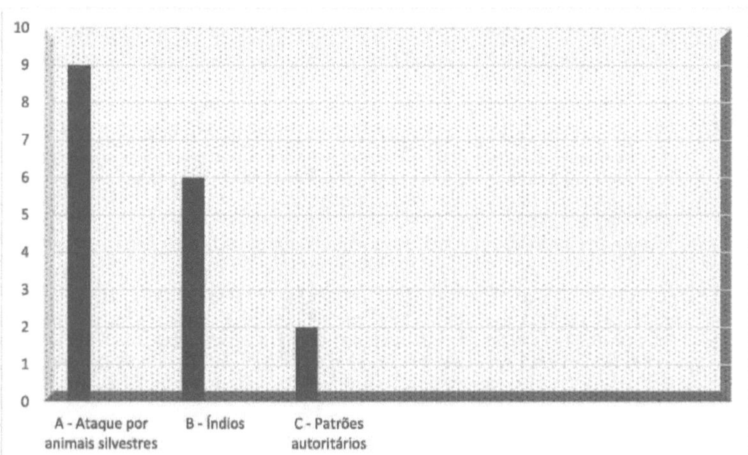

Graph 04: Distribution of categories in relation to the question: Danger of attack on rubber plantations

Source: Prepared by SILVA, M. L. S based on interviews conducted in February 2017.

Graph 04 stands out for category A -Attack by wild animals-, revealing one of the greatest dangers faced daily by rubber tappers, with an emphasis on Indians as well.

CATEGORY 04 - A

DSC4 A: Attack by wild animals

> It was very dangerous, there were lots of jaguars to attack you in the forest, it was a terrible sacrifice; My daughter died in childbirth there in the seringai, because she was bitten by a snake, and when she was born the next day, she had a baby, but she didn't survive, and because she haemorrhaged, when the snake bit her, there was nothing, she didn't feel anything, but later...., there in the seringai; There were jaguars, Ah, there were, in the good centres there were, ne; Once it was daytime, right around that time (around 10am), close to home, then I came across a jaguar, I didn't carry a shotgun because I'd left it at home, I carried a little knife about this big, the beast was roaring, like
>
> From here, this house would look at me and make such an ugly face! I thrashed about, shouting, "No bum! The more I shouted, the more I said: boy, I'm already lost for everything, only God will deliver me, so I took the little knife and cut into it. Then she ran and jumped into the woods. I remember it well: either one or the other! But I was brave at that time, I used to walk, I used to cut the whole road at night, the whole way; I don't know how I'm still alive today! I don't know how I'm still alive! because when it was night, it used to prowl all night, the jaguar used to prowl all night, then when it was dawn and we got up to go on the road, then I'd be scared to death! because they say, the people there, that jaguars are very scary, big-busted women. Ah, nega véa, I escaped with a razor's edge, because we spent a lot of time on the rubber plantation and I've been crazy all my life; dying from snake bites was difficult; there were many brave jaguars. I even escaped the jaguar twice (laughs). We used to hear that people who raised pigs would attack the pigs; Onça, vigiii, I killed about four of them, but I was always the one who killed her, she never killed me, (laughs). Where I worked for the last four years cutting, a young boy who worked with us, he was wounded by a snake, the jararaca-açú. They say that when it doesn't kill, it kills. He got sick in the leg. (CSD elaborated from statements sd1; sd2; sd3; sd4; sd5; sd6; sd7; sd9; sd10)

These speeches show that it was a very dangerous time, due to the large number of wild animals that attacked, especially jaguars, which put the lives of these rubber tappers at risk every day when they had to work on the rubber roads. These are stories of "escaping by the razor's edge", of "I don't know how I'm alive today", revealing the countless dangers they were forced to face in the rubber plantations.

DSC4 B: The Indians

> Vigii... we were in the woods a lot, hunting. They lived in the forest, but they were tame, not wild. There were Indians living there, but tame Indians; Indians at that time, I saw them like this, I even saw them passing by in canoes, paddling only on one side. The Indians at that time liked to sell the game they killed. The Indians would ask for our women's skirts. They looked and asked. They also asked for work. Only they said: caboco corta mato com a mão! And they demanded good pay, better than the white man's (laughs).
>
> Then this Batista went to attack a maloca on the banks of the river where I lived, Liberdade, there were three boys, arigó, strong, then they cut (syringe) for a year and saw no trace of Indians at all. Then one day, on a Sunday, one of them said, boy, would you like to take an abacaba for us to drink with roast meat? Then one of them said: "Boy, I'm not going to," and the other said: "I'm not going to either." Then he said: "Well, one of you get the wood, the other get the water and I'll get the abacaba. So he went to get the abacaba, took out a big sack of burlap full of abacaba and it cost him a lot, all his valour! Then when he came in with a load on his back, he went into the kitchen, and the sight took his breath away, he saw the strips, figs in one corner, bofe in another, pierced like this in the mocotó, he took out the strips and the spines, like one takes out a turd..., the Indians had done this. Then he got up and went to knock on the shed, when he arrived he fell backwards, dead tired! It was close, it was an hour's journey, and then they asked him, to die, to put perfume on his breath, because at that time there was no alcohol, right, until I realised, I'll tell you the story. (DSC elaborated from statements sd2; sd6; sd7; sd8; sd9; sd10)

Although many of the parents of the people interviewed had experienced tense moments when peace between Indians and non-Indians was still a utopia during the first rubber boom, during the second rubber boom a certain harmony was revealed between them, as they reported: there were no more "angry" Indians who attacked people, but they were already interacting with the rubber tappers, either by trading their hunts or asking for jobs and clothes.

DSC4 C: Authoritarian bosses

> Ah nega veia, when my mum lived in Seringai, they ran away from there, they ran away so they wouldn't die. In Liberdade, they ran away from Liberdade. Mum used to tell me that at that time, when someone had a balance, they would have them killed. When you had a good account, the one who was doing well, who worked hard, who had a good account, they would have you killed. My grandfather ran away from there with his family, Vo Luiz and Vo Isabel. Well, that's when they came here.
>
> The bosses took the ammunition and the rifle, if we wanted everything we bought it, everything, everything. The rubber you made, you didn't have the right to leave a rubber in the smoker, she didn't want us to leave a rubber at home, that's how it was.
>
> The bosses at that time didn't trust us, you'd make a load of rubber and hand it over to the boss, then it was difficult for anyone to take the balance, it was only debits, all the time, it was like that. A lot of people got sick at work. My grandfather used to run around here, he was one of Mâncio Lima's henchmen. He ran after the Indians in the forest to kill them, he was told to do it or he'd die. When I was a child, I got to know Véi Mâncio, but he fell ill and died, nobody knows what, up in Paraná do Pentencoste, up there. They killed a lot of people!" (CSD based on statements sd4; sd5)

This discourse reveals a time during the first wave of migration in this region, when the bosses were judges, who saw themselves as the owners of their employees, of the thousands of Indians who were expropriated, decimated and seen as animals. Times that were still lived by these people's parents.

Question 04 *"Danger of attack in the rubber plantations"* reveals that the abundance of fauna in this region put rubber tappers' lives at risk, which was the main reason for attacks when it came to working and living. It was a job that was not only painful because of the long hours the work required, but also because they had to walk for several hours due to the distance between the rubber roads in the middle of the closed forest and the proximity to the natural habitat of these animals. These were dangerous times for these workers, whether they were being attacked by irrational or rational animals.

Question 05: *Were there many diseases in the rubber plantations? How did they cure them?*

The aim of this question is to find out what these rubber tappers did in the event of illness, and whether there were many. What medicines they used to cure illnesses in the rubber plantations. Six categories were found in the speeches on this question, as shown in Graph 05.

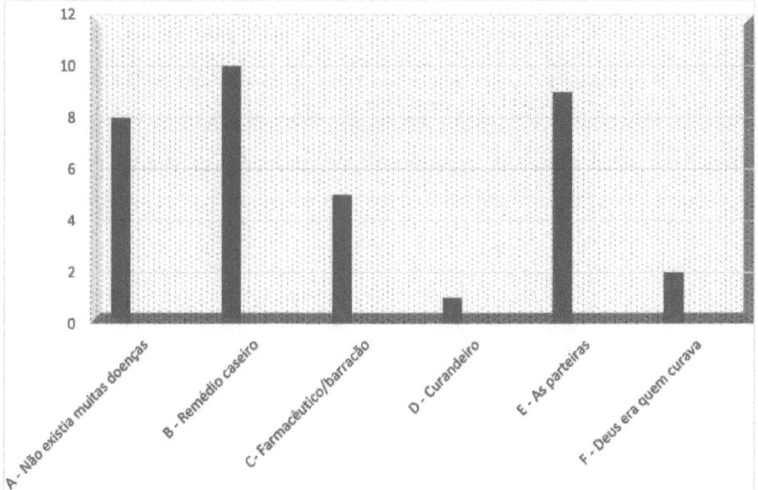

Graph 05: Distribution of categories *in relation* to the question: *Were there many diseases in the rubber plantations? How did they cure them?*

Source: Prepared by SILVA, M. L. S based on interviews conducted in February 2017.

Graph 05 highlights the categories "home remedies" and "the midwives" as the most commonly used resources within the rubber plantations, despite there not being many types of illness at the time.

CATEGORY 05 - A

CSD5 A: there weren't many diseases

> No! It was healthy; No, at that time, there was no such thing as a disease; the disease that existed was malaria, which they called cezão, the name was cezão; Tea. No-one caught the

77

flu, nowadays there's so much illness in my family. No. No disease! Thank God! In those days, we only had cesão and malaria. I never caught it, thank God, I never knew what malaria was. Most of the cures were home remedies for cesão, because there was no doctor, there was nothing. Note that I had eight men and nine women, none of whom went to hospital, I never even had an examination, no, my children were all born at home, I had two twin births. The midwives who delivered them, I never took a child like that to hospital, never, the medicine I asked them to take was mioral (melhorai), at that time there was mioral, today there isn't any more, it was calcium for me when they were piquininim, it was calcium, it was magnesium, meracilin. (DCS elaborated from statements sd1; sd2; sd3; sd4; sd5; sd6; sd8; sd9)

It can be seen in these accounts that there weren't as many diseases as there are today. The many rubber tappers and their children who were born in the rubber plantations, without medical help, relied on medicinal herbs or the shed.

CSD5 B: home remedy

We cured it with tea from the forest! We also used home-made remedies because there was nowhere to buy them, when we had a headache (and there were no pills), we took it to get over the fever, the remedy was black pepper tea, to take to get over the fever…it was, it was; Home-made remedies. Tea.

I used to cure their bugs, you know? when the bug fell, I'd scrape the paxiuba, shave it very finely, clean it, that pritume, cure that spoonful of powder, then put it in the cloth and sieve it, then when I went to bathe it, I'd put it on top of his bug, when it was clean. He cured everything. I made a bracelet straight away, I raised everything and I never went to a doctor's, a foot bath, when I could afford it, because that was a time of poverty, right? (DSC elaborated from statements sd1; sd2; sd3; sd4; sd5; sd6; sd7; sd8; sd9; sd10)

Because they were far from everything and everyone, home remedies were the main way to cure illnesses. Passed down from generation to generation, the teachings regarding the knowledge, use and handling of these herbs still permeate the memory of these people to this day, who describe the step-by-step process in detail.

DSC5 C: Pharmacist/shed

Then when they got sick with cezão, they'd get a fever, but with a little medicine they'd be fine. It was a pill, not today's pill, it was a pill! It was methotrexate, it was aralém, then you'd take two or three pills and you'd be fine. Even if I had a disease, I had a caesarean section. To cure it, it was a few pills like that. The pharmacist would take them to the shed. (DCS based on statements sd1; sd6; sd7; sd9; sd10)

These accounts show that some rubber plantations also had pharmacists. But it was Barracão that provided assistance with pills for diseases characteristic of the time, as was the case with the cession.

DSC5 D: healer

When a child fell ill, sometimes with a broken heart, they would call a healer, he would cure them, then he would give them a chazim, the medicine was homemade medicine, everything was medicine that we made and used, there was no medicine like there is nowadays. (DCS based on the statements of sd 1).

It was also common to turn to healers when they fell ill, especially when children were sick, a practice that is still common in the riverside regions.

DSC5 E: the midwives

The midwives who delivered babies, I never took a child like that to hospital, never; when a woman gave birth it was with the midwives. There was one who lived two hours away and

These accounts reveal the voluntary help provided by midwives in this region. With no
doctors, it was up to these women to walk for hours and hours, even at night, to help their partner
who was expecting a baby.

DSC5 F: God was the healer

These accounts reveal the faith of those who had no specialised medical or hospital
resources. In the event of illness or other unforeseen circumstances, when other resources didn't
work, all they had to rely on was Divine protection.

Question 05 *"Were there many diseases in the rubber plantations? How did you cure them?"*
reveals that cezão, a typical disease of the time, was the most common. Although there weren't so
many types of illnesses, when they were affected by them or by wounds, the forest itself provided
the remedy through the use of herbal teas, lambedôs, among others. In this case, the popular
knowledge of the elderly, passed down from generation to generation, came into play; the faith
exercised through the beliefs and prayers of the benzedeiras and the extraordinary empirical
knowledge of the midwives, who worked without any scientific study of medicine and without much
working equipment. This process reveals how much mastery and intimacy these people had over
the forest, their space. The scarcity of adequate medical care made them doctors of the forest. These
values, sometimes despised by modern medicine, were the great "mother" that nurtured these
thousands of rubber tappers in the remote forests.

Question 06: *Situation of women and children within the rubber plantations*

The idea here is to describe the situation of the women and children of these rubber tappers
and what they did to help with the work in the rubber plantations in order to survive.03 categories
were found in the speeches shown in graph 06:

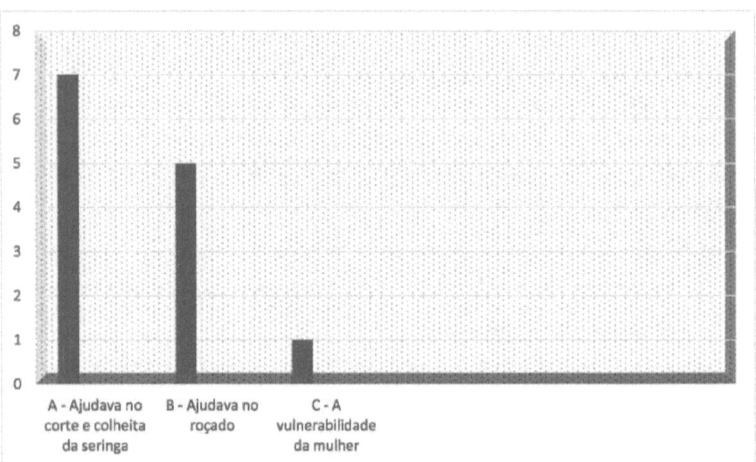

Graph 06: Distribution of categories in relation to the question: *Situation of women and children within the rubber plantations*

Source: Prepared by SILVA, M. L. S based on interviews conducted in February 2017.

Graph 06 stands out for category A "Helped cut and harvest rubber", revealing that the heavy work in the rubber plantations was not limited to men alone, but when necessary it was the women and their children who helped these rubber tappers cut and harvest rubber.

CATEGORY 06 - A

DSC6 A: helped with the cutting and harvesting of syringes

> I went once, twice, to collect it! To collect milk. It was far away, very far away; I used to carry the syringe, with my son, my daughter..., with my father, we'd go to cut with my father too, to collect the bowl ne; my father died when I was a child, I didn't know him, so I'd been working since I was single ne, with my mother, with a cousin who brought us up, it was all me, it was..., I managed. My mum was ill, I was the one who put up with my mum. I had no father, no brother, no one. Then I went to work, cutting rubber! I cut syringes for seven years. When I got married, I cut rubber, my husband cut rubber; then I was so small, then I only cut downhill, downhill is where you reach, well, then I cut and he harvested because at certain times of the day he would get up to go to the road. I had a hard time when I was cutting syringes. My father cut and I harvested, at that time I was small, bags of milk weighed a lot, I thought it was bad as hell, but I had to. (DCS based on statements sd1; sd2; sd3; sd5; sd6; sd8; sd10).

These speeches show that women also took part in the hard labour. In addition to domestic chores, they were the ones who helped with the harvest and even cutting the rubber when their father, mother or husband was absent. When they fell ill, alone or with their children, they became responsible for huge roads of syringes so as not to miss out on their obligations to the Barracão and the family's livelihood. These children have already grown up participating in this world, sometimes just accompanying their father and mother, sometimes being forced to cut, especially when their parents were ill.

DSC6"B: Helped in the fields

> I also helped plant maize, manioc, rice... I got tired of helping, both at my father's house and here with my husband, I used to go to the fields too; I used to plant more with the boys, I used to plant, I used to help in the fields, in the rubber plantation we do everything. The man goes to the rubber plantation and the woman works the fields; I have a daughter who still works the fields today, even if she's ill, poor thing The woman did everything in those days, she chopped down wood with an axe, she did everything. (DCS based on statements sd1; sd2; sd3; sdõ; sd6).

The swidden was also a place where women played a big part in providing for the family. Their children were also responsible for helping with the tasks in the fields, along with their mother, who took on the main role, "brocando roçado", which required a great deal of physical effort, since her husband was busy cutting the rubber trees.

DSC6 C: Disputes, women's vulnerability.

> I had a friend who used to tell me that at that time (the first rubber boom) there was a shortage of women in the rubber plantations, so when they learnt that a rubber tapper had a woman, they would get together, a group of ten, fifteen, and go and take the woman. If he rebelled, they would kill him there and take her. Now he says that the arrangement was that each of them would spend a month with her. The last one was obliged to stay with her. If the woman complained, she would die too. Well then, this friend of mine said that he worked with a guy called Vidal. When he came in from the road, there were 12 of them sitting on the floor, each with a 44. I came to get the woman. Then he said, are you the woman's husband? No, he's still coming, it's Vidal. When he arrived, I told him and he said oh boy, there's nothing. Then he went there happy and asked if the woman had had lunch yet. And they said: we've come to pick up your wife, and he said ah boy, you can take it. You don't have a wife, I'll get another one in Ceara. So he went over, whispered to the woman and said: you go straight ahead. He knew that the road they had travelled took a big turn. Then the woman went off crying, and when she was halfway there he said, "Boy, I'm going to kill some peccaries I left there. He went over the bridge, the first one he saw, the other one tried to run, and he killed them all. They went back to the house and he told his wife that if she wanted to stay there, she could, because he didn't want her to. He left and was never heard from again. He had a head because he already knew the story. I was a boy at the time when I heard this story. Not in my day, thank God. The woman couldn't do anything. (DSC elaborated from statement sd10)

In addition to all the work women did in the rubber plantations, they were also subjected to a situation of greater social vulnerability, a case apart from what they experienced at the time (portrayed here in the first migratory outbreak). They were treated as objects, with a total lack of respect, and they were fought over because there was a great shortage of women and there were no laws to protect them. Many went through situations of having to "serve" several men, when they stole them from their husbands.

Thus, question 06 - *Situation of women and children in the rubber plantations* - shows that as far as her role as a worker was concerned, she was her husband's "right-hand man". Her work went far beyond domestic chores. In various situations, she took the lead in cutting the rubber, in the swidden, exploiting her leadership and skill when it came to taking on her family responsibilities. She was always warlike and intrepid in the face of the surprises that life brought her, even in the face of the high social vulnerability they experienced at the time.

Question 07: Were you *able to study?*

The idea here is to find out whether these rubber tappers, their children and their wives have had access to education, even if only afterwards, in the present day.This question found four categories in the speeches shown in graph 07.

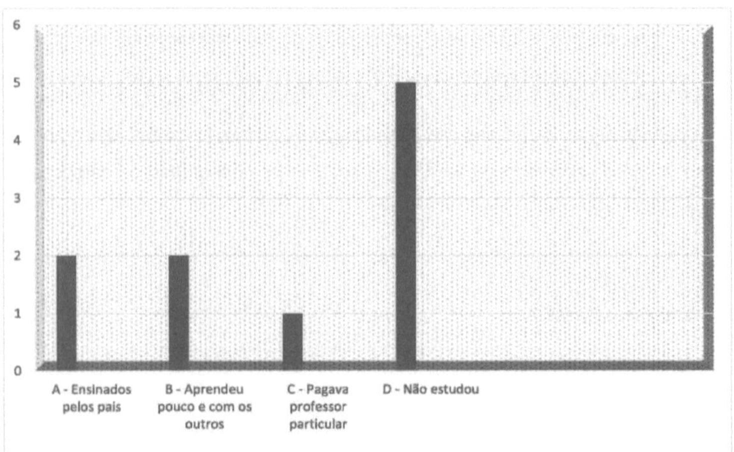

Graph 07: Distribution of categories for the question: *Did you manage to study?*

Source: Prepared by SILVA, M. L. S based on interviews conducted in February 2017.

Graph 07 highlights category D - Didn't study, revealing that the majority of these individuals had no access to education within the rubber plantations.

CATEGORY 07 - A

DSC7 A: Taught by parents

> I was only able to study because Dad taught me. My mum knew how to read and write and so did he; only at home, my dad knew how to sign his name very little. (DCS based on statements sd1; sd9)

With no schools, in difficult-to-access rubber plantations, these children often relied on the teaching of their parents, those who arrived here with some knowledge. It was just the basics, like signing their name, because there was no time available due to the long working hours their parents had cutting rubber.

DSC7 B: He learnt little and from others

> Ah, I had no way of studying..., no... I studied like this, with those who knew, I asked for lessons from those who knew, I'm not a graduate, I don't have a class. I learnt to read, I learnt to count, I learnt to write a letter, everything, but..., it's all in my head. Today I don't do anything anymore, at the age I am now, my eyesight can't write anymore. (DCS based on statements sd3; sd8)

The lack of schools meant that the children of these rubber tappers, and when they had the interest, time and inclination, looked for other ways, such as using those who knew. However, it was always the basics, just "signing your name".

DSC7 C: I paid a private teacher

> Ah, I didn't study much. I studied in a seringai, then I studied because I paid for two private teachers, but it wasn't much. Sd10

In rare cases, we still find those who were willing to pay private tutors to study. But the characteristic of this study, and like the others, was also "I studied little", which for them at that time was already a reason to be something.

DSC7 D: Didn't study

> My father could read and write. I didn't know, I never studied. I'm not going to say I didn't, because I never went to school. My school was to work in the fields. There was no school, there wasn't, there wasn't; there wasn't. I was brought up like a brute in the world, without studying or anything, I don't even know anything, I know these letters, but knowing how to read, I don't know. I still had to go to some classes, after I arrived in Mâncio Lima..., well, that's when I learnt how to read.
> I said: I'm leaving because Dad won't learn to speak, so I left; Here, we were all brought up without any schooling. Dad wanted to rent a house out here, to put mum with us, to study, she didn't want to, because of the people talking; Here, we were all brought up without studying. Dad wanted to rent a house out here, to put Mum with us, to study, she didn't want to, because of the people's talk; I didn't study, my daughter. After I arrived here in Mâncio Lima, I enrolled to study, but I didn't have time, because I had my daughters to support, and at night I was napping. But I really regretted it. (DCS based on statements sd2; sd4; sdõ; sd6; sd7)

In these accounts we see that for many of these thousands of northeasterners, or their children, access to education was not possible. Whether due to lack of incentive, lack of time or even being overcome by their own fatigue from working in the forest. Others were afraid of being "talked down to" by others when it came to being alone, away from their spouse in order to give their children access to study.

In response to question 07-0 *"Did you manage to study?"*, we noticed that studying was rare and difficult. For the few who managed to study, they learned from their own parents, who taught them in their spare time, or from friends who, having some knowledge, were willing to teach them, since the few teachers in the region didn't reach everyone due to the difficult access or even the heavy workload imposed on these rubber tappers. Furthermore, it wasn't the aim of the big bosses to give these rubber tappers an education, but rather to exploit the large mass of cheap labour they had through hard and difficult work.

Question 08: *Contact with family in the north-east after coming to Acre*

The question aims to find out if, after these families migrated to Acre, they had any contact with the family they left behind in the Northeast. 03 categories were found in the speeches shown in graph 08.

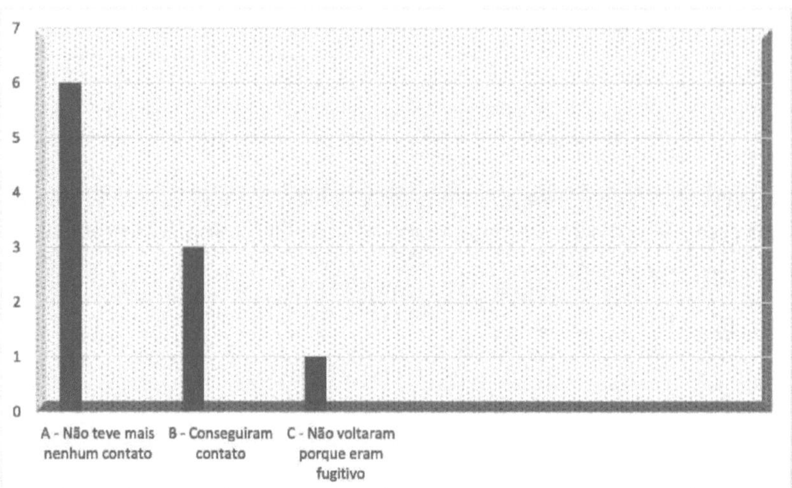

Graph 08: Distribution of categories in relation to the question: *Contact with family in the north-east after coming to Acre*

Source: Prepared by SILVA, M. L. S based on interviews conducted in February 2017.

Graph 08 highlighted category A - No further contact, demonstrating the total break in family ties left behind in the north-east.

CATEGORY 08 - A

DSC8 A: No further contact

> Not my husband! He had no contact with anyone there. He only said he had 20 brothers. Now when he came here, his mother and father had already died, it's a big family, but he never had contact with anyone... he never had contact again, he lost everything. My mum really wanted to see her family again, but she died in 1942. She died and never went back. He used to say that he had left his brother, mother and father behind; Well, all I know is that the late Joao (father-in-law) was never able to go to his hometown again, no-one ever heard from his family again, no! iiije, there are lots of relatives there, Ceará is big, ne! (DSC elaborated from statements sd1; sd2; sd3; sd4; sd5; sd7; sd8).

It can be seen in these speeches that most of those who came to cut rubber, whether conscripted or voluntarily, had no contact with their families left behind in the Northeast. They totally lost their family ties. Fathers, mothers and countless siblings were totally separated due to the great distance and lack of communication at the time.

But although most of them haven't been lucky enough to see their relatives again, some have managed the feat of reuniting with them after long decades and show how they managed to find them.

DSC8 B: They managed to get in touch

My husband died and he never had any contact with his family from Ceara. Some of them
came back. Now, after he died, two of my sons tracked down the family on the internet and
went looking for them. If he'd been alive, what a joy he wouldn't have had. He'd talk about going
to Ceara. He'd say: oh I'm not going to take my children to live in Ceará, I'm going to take my
children for a walk in Ceará, he knew he didn't want to raise his children in the drought, I want
to go there and show them my family, we'll go and come back. Now they're organising
themselves to come here to Acre to see us. There were only ten men with me, and four sisters.
I was sixty years old and I hadn't been there. He managed to get there. He managed to go
there already. About four times. But look, it's the saddest thing in the world not to know about
your family. Dad died and I dreamt at night that he was blind and separated from Mum. Then
I'd say to the people: Dad died, Mum didn't, when she died I didn't know. (silence..) I didn't,
because I really didn't; Boy, at some point, about ten years ago, my daughter worked at
Teleacre, and there she had contact with one of my father's nieces. Because in '43 my father
heard from him and gave him the address, but then lost contact. My daughter had contact with
a politician in Fortaleza, well, a politician, right? so he rang me, gave me the address, the next
day he got it, I spoke to this niece of mine. But then we lost contact again. Nobody knows him
any more. (DSC based on statements sd6; sd9; sd10)

These speeches reveal that those few who were fortunate enough to live for long years
managed to contact their families in the north-east. Helped by the advent of current technologies,
which shorten time and space, they have become powerful weapons for finding the thousands of
loved ones left behind in the north-east of Brazil. Others, however, have used the Amazon as a
refuge, unable to make contact simply because they are fugitives from justice, and so have chosen
not to risk their fate by staying "safe" in the bars of the forests.

DSC8 C: They didn't come back because they were fugitives.

These guys who came from Ceará never went back, it was because they couldn't be there,
they did a lot of wrong things, so they were forced to flee. (DSC elaborated with sd2
testimonies)

There were also those, as this story shows, who were denied access to their loved ones for
fear of returning and facing justice. These were the fugitives who came during the Second World
War

With regard to question 08 - *Contact with family in the north-east after coming to Acre* - what
was already predicted emerges: because this region is difficult to access and the most extreme
western point of the country, these rubber tappers were prevented from seeing their loved ones
again. Few of them managed to live for many years and were graced by the advent of new
information and transport technologies, making it easier to meet their loved ones. Others weren't as
lucky and to this day have been unable to get in touch again, either because they can't be found or
because they're afraid to return because they're fugitives from the law. A family link lost in the
confines of the Acre jungle.

Question 09: *Did you make a lot of money from cutting the syringe?*

The aim is to find out whether the federal government's propaganda has materialised, through
the promises that this was a land of fortune-making, of making a lot of money.

In this question, 03 categories were found in the speeches shown in graph 09, which are:

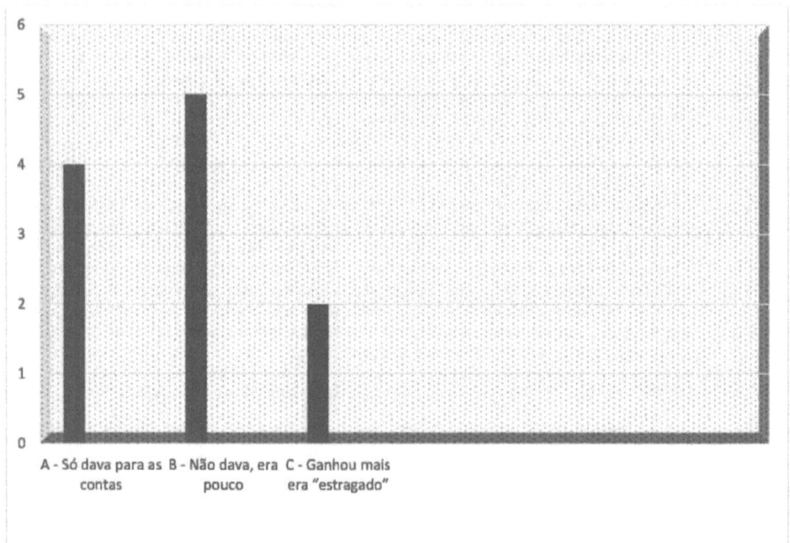

Graph 09: Distribution of categories in relation to the question: *Did you earn a lot of money from cutting rubber?*

Source: Prepared by SILVA, M. L. S based on interviews conducted in February 2017.

Graph 09 stands out for category B - It wasn't enough, it was too little, emphasising that the money earned from cutting rubber was insufficient to meet daily needs in the rubber plantations.

CATEGORY 09 - A

DSC9 A: It was only enough for the accounts

> It was just enough to pay the bills. My father, my father had my brothers, there were five children, me and Nazaré. Three of them cut rubber and Dad also worked in agriculture. There were five men and two women. We were able to earn enough to buy what we needed, to buy the little things we needed for the house.

It can be seen in these speeches that the money collected from cutting rubber was only enough to pay for the rubber tapper's expenses, breaking down what the countless national campaigns were propagating, which spoke of enrichment, a lot of money for this rubber tapper.

DSC9 B: I couldn't, it was too little

> No, when the year ended, at that time everything was very little, that was it; No. Rubber tappers at that time didn't get any money, when they made rubber they would go to the shed to buy all their necessities, then they would come home with all the goods; No. My father, the money my father earned was from rubber tapping. No. I didn't earn any, because the rubber plantations I worked on produced very little milk. Rubber tappers made a lot of money on the rubber plantations that were good for milk, they made a lot of rubber. (DSC elaborated from statements sd4; sd5; sd6; sd7; sdsdlO).

These people reveal how difficult it was for them to survive because of the little money they collected, which was not enough to meet their needs. There were also those who, even though they earned money, were unable to manage it and ended up losing everything.

DSC9 C: won more was "spoilt"

> I won, I was someone who always had a balance. I made a lot of money on the syringe, but only at that time, I got married and when I'd finished getting the balance, I'd come and spend three months here, single, I'd spend it all, in one place, in another, at parties, it was very spoilt. But when I started to build a family, I didn't let them go hungry. Boy, I earned a lot of money, but I would squander it, I would give it all away, the last money I had, I already had a family, I would travel to what, to glue, then it would run out. When I didn't have any more, I'd earn it again, ne. (DSC based on statements sd8; sd9).

These speeches reveal that those who managed to get good balances didn't manage them properly and ended up "ruining" everything they had achieved with their hard work for months.

In response to question 09 - *Did you make a lot of money from cutting the rubber tree?* - all the attractive propaganda about the northeast of Brazil, showing a land of abundant wealth and easy victory, falls to the ground. Yes, there was an abundance of money, but without a doubt the many riches exploited in this region were not destined for these rubber tappers, but for their bosses, the high powers that ruled this area. It was up to the rubber tapper to be lucky enough to get any money at all.

Question 10: *Comparing before and now*

The idea of this question is to take stock, to compare your life before and now. In this question, 03 categories were found in the speeches shown in Graph 10.

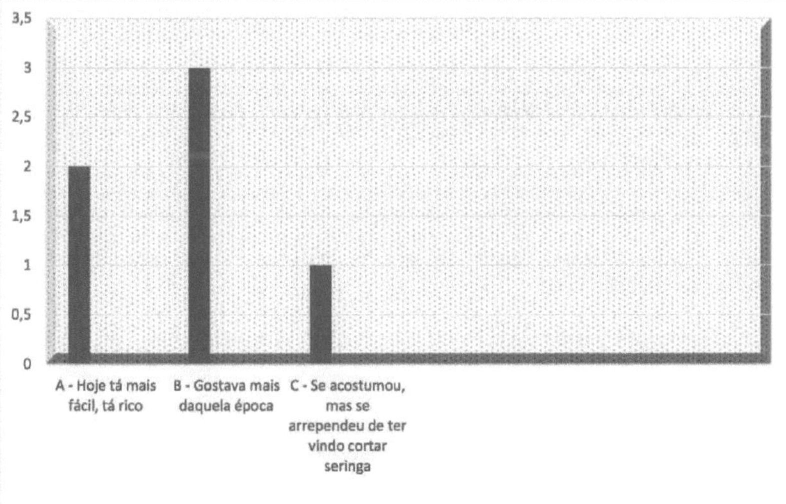

Graph 10: Distribution of categories in relation to the question: *Comparing before with now*

Source: Prepared by SILVA, M. L. S based on interviews conducted in February 2017.

Graph 10 stands out for category B - I liked that time better, demonstrating that despite all the adversities, these rubber tappers still have a greater appreciation for that time than the current one.

CATEGORY 10-A

DSC10 A: Today it's easier, it's richer

> Today they're rich! People are all rich, compared to the old days they're all rich, because in the old days you'd go a year without seeing a penny in your hand, you had nothing to live on, you had nothing to receive, it was such poverty; Today people who used to be all rubber tappers, from the fields, struggling for a living, today everyone has a car, a motorbike, a house, so everything is easy, ne. Everything, everything, is easier today. Today everyone has a car. My daughters think it's funny when I say: So everything is rich, everyone is rich today. (DSC elaborated from statements sd1; sd2)

In these speeches we can see the assessment made by these many rubber tappers. According to these narratives, today everything is easier, whether it's acquiring goods or finding a job. This ease of access to things is so immense that today it is compared to great wealth, in other words, today these people have achieved wealth. But the majority showed their appreciation for that time, despite the difficulties, and revealed why they preferred it.

CSD10 B: I liked that time better

> I liked that time, I worked a lot, but I liked it. I was poor, I've been poor all my life. I was born in poverty, I grew up in poverty, I went hungry, I went thirsty, I worked so hard, but I liked it, it was the life that God gave me, that's it, all the crises I went through, nowadays, thank God, God is recovering me, I'm not going through crises anymore. I've cut out the syringe...

> It was a lot of sacrifice, but we could live, there wasn't as much violence as there is today, only if it was something like this, an accident, a stick in the woods, a snake, but people would leave and we'd be calm. Nowadays, if a person goes anywhere, we're already thinking about it, but not back then. Good times, I thought, were ours. And everyone didn't have an education..., but everyone had an education, respect, domestic education, everyone had it; not for me. I thought it was better at the time because I was working, well off. I live sick. Now there's just one thing, for those who are healthy and want to work, today is better, because back then it was harder to sell things and today it's much easier; Today everything is easier. Now it's not for me because I'm old, I don't have the strength, I can't walk any more. I don't have the strength to pick up an object. But I miss the woods. That's what I miss, the time I used to do things. (DSC elaborated from statements sd3; sd8; sd10).

In these accounts, the subjects emphasise that, despite all the difficulties they faced, the poverty, they have a greater appreciation for that time. They describe the difference between before and now in terms of daily security, current illnesses, the loss of domestic education, values that were sacred in those days. Longing for the forest is also one of the things that brings back good memories, mainly because those were times when they could work.

There were also those who showed deep regret about their choice to come to the region. Time made him get used to the distance.

CSD10 C: He got used to it, but regretted having come to cut the syringe.

I thought it was bad at first, but then I got used to it...

I regretted it, Virgin Mary. I cried, I'd never cried before in my life. Many times, on the road, I cried my eyes out, I spent hours on coke, imagining. It's a crazy regret, isn't it? Look, I lost everything I owned there, I lost everything. When I got back to Ceara, I couldn't find anything else. Nothing of my parents' inheritance! But my sister, where are my inheritances, my sisters? Oh my fii, so-and-so got everything. He cheated so-and-so, he cheated cicrano, and belano and so-and-so, I say, but I'm not even going to talk about it, I didn't come here looking for any of that, I don't want to! If it's okay, if it's not, I'm not being silly either. This man, you know, he knelt at my feet, he was my brother, he was younger than me. He knelt here and asked for forgiveness for what he had told the judge, that they had killed me in Acre, that he was no longer alive. That's why I lost my inheritance. He kept everything, sold the farm, swapped it for a house, and the other one they took and sold, everything. I came back here because I had my family, but if I hadn't I wouldn't have come back here. (DSC based on statements sd10).

This speech reveals the regret of a man who, having been drawn by lot, chose to come to the Amazon instead of the war. Although he got used to the new situation, he never lost the desire to return to his land, to his people, one day.

Question 10 - *Comparing before with now - was* a surprise when the respondents admitted to preferring that time. This strangeness is soon backed up by reports comparing the current situation of violence such as robberies, thefts and the lack of security that the country is currently experiencing. The change in family values is also a cause of strangeness for them, as they can't get used to the normality of so many disagreements between parents and children, domestic violence, family neglect and so on. That would be the reason why, even in the face of all the immense hardship and deprivation they suffered at that time, they still opted for it.

4.2 Discussions

The end of the 19th century brought to the Amazon a historical event never seen before in this region. The large-scale occupation through the displacement of thousands of people, almost entirely from the north-east of Brazil, for what is known as the rubber boom, thus characterising a period in which rubber had become one of the country's main economies, condensed an important stage in the territorial occupation of the Amazon, whether by spontaneous act, by enticement (convinced by government propaganda) or recruitment to make up the specific workforce for cutting rubber, the "white gold" of the time.

What the great works tell us is that this period was the golden age of rubber, and that the emperors of this region were rich. A "beautiful" story full of memories for those who managed to make a fortune from this source.

However, it is not common to find works that talk about this time from the point of view of those who came to make up the labour force, the rubber tappers. Why not give them a voice? What is the reason for the lack of interest in listening to someone who was intimately involved in this whole historical process? These questions sparked this research, to look at this new angle and discover what is hidden in the "cracks" of history and that the new and great revelations often go against the great stories that books tell us. The idea is not to reveal a result, but to show that there are other stories.

But what do these stories reveal? Everything that I already knew, and I say this because I am also a descendant of these people, with great honour, and because I spent my childhood listening to the stories of my ancestors' experiences in the rubber plantations. That's why I'm so keen to give these people a voice through their stories,

> When a person recounts their memories, they transmit emotions and experiences that can and should be shared, transformed into experiences so that they can escape oblivion. The moment an interview takes place, the interviewee finds an interlocutor with whom they can exchange impressions about the life that goes on around them. It is a moment in which memories are ordered in order to give meaning, with the help of imagination or nostalgia, to the experience of the subject narrating their story. (SANTOS, 2005, p.3)

Therefore, these narratives are important historical sources for finding out details that are not found in official documents and so that we can get to know the lived reality of these people. They can contain a lot of specific information, helping to explain many questions. Furthermore, this is a fundamental resource for making a given moment interpretable through narratives of everyday experiences, events, memories, dreams and desires.

A curious aspect of this migration process is the fact that the story is told only about the rubber soldiers, making women invisible in the region during this period. Perhaps this is why there is little research on them, since it was only the men who were recorded as coming to the Amazon. But according to the accounts in this work, they were fundamental to survival and were extraordinarily useful during this period:

> [....] at that time, (the first rubber boom) there was a shortage of women in the rubber plantations, so when they knew that a rubber tapper had a woman, they would get together, a group of ten, fifteen, and go and take the woman. If he rebelled, they would kill him there and take her. Now he says that the arrangement was that each of them would spend a month with her. The last one was obliged to stay with her. If the woman complained, she would die too [....].
>
> It was far away, very far away; I used to go to the syringe, with my son, my daughter [....], with my father, we'd go to cut with my father too, to harvest to collect the bowl, right; my father died when I was a child, I didn't know him, so I'd been working since I was single, right, with my mother, with a cousin who brought us up, everything was me, it was..., I managed. My mum was ill, I was the one who put up with my mum. I had no father, no brother, no one. Then I went to work, cutting rubber! I cut syringes for seven years. When I got married, I cut syringes, my husband cut syringes [....].
>
> I also helped plant maize, manioc, rice... I got tired of helping out, both at my father's house and here with my husband, I used to go to the fields too; I used to plant more with the boys, I used to plant, I used to help in the fields, in the rubber plantation we do everything, right? The man goes to the rubber plantation and the woman digs the fields; I have a daughter who still digs the fields today, even if she's sick, poor thing The woman did everything back then, she chopped down wood, with an axe, she did everything [....] (Interview granted in February 2017)

These accounts show how difficult the situation was for women, where the oppressive patriarchy ruled and dictated the rules. With no greater prospects, these women were treated as a product of personal conquest to serve men's desires. In addition, the role of rubber tapper and farmer was also very well exercised by them, although they were never credited with it. Men's labour was easily shared with women, especially at crucial moments in their lives, such as when they were ill. That's why those who were born in the region, in other words, the descendants of these migrants, were trained from a young age in hard labour, along with the family. It was a portrait of a harsh, painful, difficult and severe time.

But the question remains: why do most of them still miss that time when they say: "I liked it then, I worked a lot, but I liked it" (DSC10 B, p.118). Wouldn't that be contradictory? Perhaps this question can be answered through the ideas of Halbwachs (1990):

> It is certainly inevitable that the transformations of a city and the simple demolition of a house will upset some people's habits, disturb them and disconcert them [....] were part of their small universe and whose memories are linked to these images, now erased forever, they feel that a whole part of themselves is dead with these things and regret that they didn't last for as long as they had left to live. (HALBWACHS, 1990, p.37).

This idea would justify their nostalgia for a time that portrays a space of longing, belonging and attachment and, despite defining them as difficult times, they still opted for them: "But I miss the forest. That's what I miss, the time of doing things" (DSC10 B, p.119). Added to this is the lack of acceptance of the new social and family standards in force today:

> It was a lot of sacrifice, but we could live, there wasn't as much violence as there is today. Nowadays, if a person goes out, wherever they go, we're already thinking, but not back then, people went out and we were calm.

> Good times, I thought, were ours. And everyone didn't have any schooling..., but everyone had an education, respect, domestic upbringing, right? (DSC10 B, p.119).

Despite the enormous difficulties, the majority still prefer that time. In addition, this whole process reveals that these rubber tappers did not receive the proper assistance that the agreements made guaranteed, for example, medical and health care for the rubber soldiers or the supply and increase of food production, the acquisition of goods, their transport and the formation of stocks. Benchimol (2010, p.280) reveals that "it was common for supplies to be unavailable for transport upriver... and for foodstuffs to rot in Belém and Manaus because they arrived there when the upper rivers were dry. The result: a year of deprivation and no production." This is the contradiction that these soldiers experienced under the promise of prosperity in the remote North.

In this context, a new question arises: What would lead thousands of people to submit to such a process, as was the case during the second rubber boom in the 1940s, even though they already had some accounts of the difficult reality of the rubber tapping system? For this question, we will base ourselves on Bakhtin's (1999) statements when he analyses the sign as a category intrinsically linked to ideology. According to him, everything that is ideological has a meaning and refers to something outside of itself, "everything that is ideological is a sign and without signs there is no ideology" (BAKHTIN, 1999, p. 32). The word would be an ideological phenomenon par excellence because every word is absorbed by its function as a sign, created for an ideological function. The privileged material of communication in everyday life is the word. But what does this have to do with our subject?

During the Second World War, which began in 1939, it was common in the north-east of Brazil to come across several posters scattered around the towns motivating men to go to the Amazon to cut rubber. The strategy began with the *Washington Agreement* between Brazil and the USA to make up for the lack of rubber in the war camps, in which the Brazilian government undertook to increase the activities of the service of recruiting, sending and placing workers in the rubber plantations of the Amazon. These posters, very well designed and idealised by Jean-Pierre Chabloz, a Swiss painter hired by the Brazilian government to idealise a bountiful, prosperous Amazon full of the possibility of rapid enrichment, appear as an ideological sign, as stated by Bakhtin (1999). As an example, figure 15 portrays this sign, as it is designed with the ideology of persuading the northeasterners to go and cut rubber in the Amazon.

Figure 16: Poster designed by Swiss painter Jean-Pierre Chabloz in the 1940s

Source: SANTANA, M., 2012

Figure 16 shows a very "tidy" rubber plantation with a pretty house surrounded by domestic animals, plantations, rubber trees all around it and the rubber tapper easily harvesting latex in the daylight, with trees next to each other on fertile land. However, the poster does not reflect the reality of the native rubber plantations of the Amazon, since the layout of the trees is unrealistic. The short, regular distance between the rubber trees was much more in keeping with the reality of the rubber plantations cultivated in Southeast Asia. The physical characteristics of the Amazonian rubber plantations were very different to those in the poster, full of tortuous, bumpy and dangerous paths (MORAES, 2010).

The reality of the poster was very different from that of the Amazonian rubber plantations, given the existence of all kinds of obstacles found within a rubber plantation, as we see in the accounts collected: "suffering in the forest, cutting rubber, hungry, going out".

from home at midnight to cut rubber... The rubber tappers carried the rubber on their backs for hours and hours to get home to their little huts" (DSC2 A, p.89). It can be seen that they use the expression "barraquinha", "leaving home at midnight", "carrying rubber on their backs for hours and hours", contradicting the image on the poster. The situation was so contradictory that the figures themselves prove the chaos. According to Moraes (2010), an estimated fifteen to twenty-five thousand workers died in the heart of the forest, victims of the inhumane working conditions to which they were subjected, as well as malnutrition, malaria and yellow fever.

The purpose of the poster under analysis is clear: to attract labour, through a painting planned in advance in order to arouse the potential migrant's desire for the "facilities" he would find in the new job. This demonstrates the power of influence and thinking of a government through an image aimed at a specific social group: the Northeasterners.

Analysing these contradictions in Bakhtin (1999), he reveals that every ideological sign confronts contradictory indices of value. The sign would be like an arena for class struggle in which the dominant party seeks to take the monovalent sign, that is, with only one argument, which would be a new life full of abundance, but deep down this ideological sign has two facets. This would explain the contradiction between the poster designed by Chabloz and the reality experienced by the rubber tappers. On the one hand, the government persuading people to migrate to a region with the argument of a "land of plenty"; on the other, the true reality reported by the rubber tappers who experienced the contradiction of the advertisement and the other facet revealed, the use of cheap labour for unhealthy and hard work.

There was also a lot of "help" from the radio, newspapers and daily appeals at the time when these future rubber tappers were encouraged to make their biggest and best contribution to Brazil's victory, fulfilling their duty as sons of Brazil, the "soldiers of freedom". It was a call to obey with energy and goodwill so that later they could look back on their past with pride and their heads held high. (NASCIMENTO SILVA, 2000).

According to Santana (2012), this discourse penetrated homes on a daily basis, persuading entire families to migrate to the "new world" of "comfort and dignity". They were adverts urging citizens to be brave, to not shirk their duty as "soldiers of freedom" and to be sure of the pride they would feel when they looked back and saw what they had done out of love for Brazil. The speech evokes Brazilian patriotism, fighting for something that would be in obedience to a duty to their country. The effects were enormous and thousands of them saw the possibility of something good for them.

Pierre Bourdieu also makes a detailed analysis of the power of the symbol and states that "symbolic power is a power to construct reality that tends to establish an order" (BOURDIEU, 1989, p. 9). He emphasises that the dominant ideology intentionally hides the conflicts and interests of classes and the means of production, which is why words are loaded with symbolic production as a structure of domination.

We can see that the symbolic power that both authors talk about is a power that constructs a reality, that establishes an order, a meaning or even conformism, which I mentioned earlier about these northeasterners believing that migrating was the "best" thing for them at the time. This is a strong instrument of domination that even affects the division of labour and, therefore, the division of classes.

All this imagery created by the government media is clear in the accounts collected by Benchimol (1977) when he translates the hopes of these rubber tappers: "I'm going to be a happy man, and if God helps me I'm going to get rich in rubber tapping"; "I'm going to Acre, but if God gives me life and resources, I think I'll still go back there"; "My destination is Acre, I don't know when I'll go back, but I intend to"; "I think I'll go back in about six years".

In addition to this, it is also possible to add the fact that shortly before the northeastern migration, the Vargas government had manifested itself in promoting the occupation of the North, the "occupation of empty spaces". Vargas' colonisation programme, known as the "march to the west", was intended to expand the frontiers by settling the interior, bringing the whole territory together and controlling the population. To this end, he promoted the creation of housing colonies in various states, including Goiás, Mato Grosso, Pará, Maranhão and Amazonas (CASSIANO, 2002).

The government's main target was the northeasterners who lived in regions with little economic development and who were going through another major drought at the time. In these processes, significant contingents of migrants arrived in the Amazon to cut rubber (NASCIMENTO SILVA, 2000).

The graphs generated through the DCS of the interviews show that in the Northeast the Amazon region was presented as rich, prosperous and a guarantee of a new life for those who bet on it. For those who didn't, during the Second World War, they were "lucky" enough to be selected to work as "rubber soldiers" in the Amazonian rubber plantations with strong and effective instruments of persuasion that it was "the best thing to do for the country". The tiring and dangerous journey didn't seem at all like what was advertised on those posters. When they arrived at the rubber plantations, reality clashed with these advertisements: reality was very difficult! They experienced all kinds of dangers (attacks by wild animals, by Indians, and also by dictator bosses). The forest and the river were their main sources of survival in the search for food. As for medical care, they also relied on the forest for help with home remedies. Midwives played an extraordinary role at that time in the dense forests. In fact, gender was a major issue, and even in the face of their vulnerability, they stood out for their enormous role of empowerment when faced with the health problems of the leaders of their households, showing the utopia of female fragility. Their work went beyond that of the household and they took on the swidden and the cutting of the rubber trees when necessary.

With no medical care, no schooling and no contact with their families, they were unable to realise their "new life" project of earning money from cutting syringes as promised. Many of them have become accustomed to their new life and say they like it better then than now, due to the conditions in which they live today. With no one to complain to or turn to, isolated and far away from everything and everyone, their dreams were cut short. Many died without memory, without history, without recognition. They left their descendants the legacy of a bitter experience.

The data collected here are reserves of memories, growing every moment, of the totality of their lived experiences. It is precisely because of them that the present body is able to relate to the past. It is not only related to these individuals, but their interpersonal reality with social institutions, family, social class, church, etc. (BOSI, 1983).

It is through these sumptuous accounts that we can add more to the story. Uncut, uncensored, but revealing that it is in these gaps that the truest and most significant stories are to be found, because they were really lived, experienced day after day. It is these experiences that make up the history of each individual with a weight of glory.

CHAPTER 5

FINAL CONSIDERATIONS

The situation of the rubber tappers during the two rubber economies, especially those shown in the interviews in this paper, reveals a political scheme planned for this region, where it is clear that those with power in their hands subjugated those who had little information. Thrown to their fate by those who recruited them, they have over time become portraits of invisibilised peoples.

Analysing the processes, the reasons that drove a given migration in a given place and time, through the accounts of people who lived through each moment, giving voice, promoting reflection, is not only being able to rewrite history, but also being able to recover a past time and make it present. It's being able to look at it from a new angle, revealing facts other than those told by the sources we usually look for: historical books. Being able to relive this time is being able to make new discoveries by valuing the voice of these once invisible subjects, not worrying about explaining everything, but raising new questions. This was our aim in this work through the former rubber tappers living in Acre.

The rush to cut rubber in the Amazon in the first *boom of* this economy was strongly driven by the external industrial and economic transformations that were taking place at this time, primarily in the USA, directly affecting the lives of thousands of Brazilians, especially those in the north-east of Brazil. The second phase of this economy took place during the Second World War. Unlike the first, this was a desperate attempt to revive what had once been considered the country's main economy, latex extraction in the Amazon region. In truth, it was a dramatic attempt to turn around an economy that had been suffering from failure for decades. Through agreements signed between Brazil and the United States, Brazil saw the possibility of reviving rubber production once again, guaranteed by the technical and financial support it received from the Americans. It's no coincidence that this battle featured armed strategies from the top ministerial echelons of both governments, who set their policies and actions by setting up a logistical-institutional apparatus that was huge for the time.

What was the main target of this detailed and articulated policy? The north-east of Brazil, which, ironically, like the first wave of migration, was suffering another major drought. They were bombarded with heavy advertising from the federal government encouraging migration to the Amazon. Advertising methods with persuasive discourse emphasising the Amazon as a land of plenty, a land of victory, a land of white gold. Propaganda with great ideological influence, but with no connection to the reality that awaited these workers. There was also the call to be a "rubber soldier", out of love and reverence for the beloved country. Everything was thought up, planned and organised by the Brazilian government. The idea was to introduce into their minds a certainty that going to the Amazon would be the best thing to do in order to have a different life from the one back in the northeastern hinterland, full of money, invoices and victories. A new life! The result was a rush of thousands who came in search of their "pot of gold".

Neither of these two waves of migration brought about the realisation of the dreams implanted by the government through advertising policies, except for those who held all the power, the rubber plantation owners. Nobody became rich, nobody's life was transformed by the wealth of rubber. It was all utopia. The dream of wealth and a return to a better life was smoke, an illusion. Quite the opposite, they were subjected to painful, dangerous work and left to the harsh, inhumane structure of exploitative labour. Prevented from escaping from such a situation by the debts incurred in the rubber plantation, by isolation and loneliness, the rubber tapper was apparently free, but in reality he was a prisoner of the rigid system imposed by dictatorial bosses within the rubber plantations. For many, this was a path of no return, as their loved ones left behind in the north-east would never come back. Thousands died abandoned in the Amazon in the most varied and negligent situations. Rare are those who have managed to see their loved ones again. Even rarer are those who still survive to tell us the true story. Rare!

Based on these analyses, the question arises: were these rubber tappers the losers of this process? In the light of this context, we say, Never! On the contrary, they were more than victorious and rich, not in money, as they promised, but in other values, experiences that no money can buy. And there were many of them: they showed that they were strong, that they were fearless in the face of danger, whether from animals or from their violent bosses, from the typical diseases that afflicted them, from the difficulty of getting from one rubber tapper to another, from the hunger they endured, from the clothes they didn't have to wear or the education they didn't have. In fact, the latter proved not to be the most important thing, because the knowledge they had was and is the most important thing in every situation. As for nostalgia, well, that was kept delicately in the confines of the mind.

Even though they were victims of human rights violations, and in some situations lived under a regime of slavery, they dedicated all their energies to the battle for rubber in the Amazon, without fear. This Amazon of rubber tappers, northeasterners, rubber, rivers, forests, experiences was the scene of great moments that reconfigured this space, resulting in a new way of life through the "heroes of the rubber tree". They were the great and true heroes in the conquest of what we now know as the state of Acre. Many gave their lives in this bloody battle. They were forerunners in this process, adding new history and new cultural values to the region. As a result of all this, they gave rise to a new place, shown by their work, way of life, experiences, suffering and hopes that never ended. As a result, they have passed on to their descendants that they are capable in the midst of all adversity. The incessant search for better living conditions was for those who never gave up hope of one day winning the long-awaited "Eldorado".

Their names are not printed on any monument, book or historical building, nor are they on any list of heroes. But it was by listening to their stories that we discovered that they were more than heroes, they became emperors of knowledge, rich in courage, in overcoming, who developed their own methods of survival. They were the most important soldiers, and contrary to what official history elects, these were undoubtedly the true heroes of the Amazon. They arrived, settled in the new space and formed the basis of what is now known as Acre society,

This whole context reveals the scope outlined in the objectives of this work: the narratives of these subjects deconstruct every fallacy implanted by the propaganda of the time. The region was very different from what was preached on the posters. The days were more than difficult! However, they absorbed the new space as their own. They became so intimate with it that even though they revealed all the limitations they had experienced, they still opted for the place "of the forest", of the rubber plantations, of nostalgia. We could think of a contradiction, but this is supported by Tuan's (1983) idea that space becomes place (the aggregation of feelings, meanings and sensitivities embedded in it). This would explain their desire for the rubber plantation because of the experiences and emotions they absorbed there.

Finally, we would like to emphasise that the purpose of this research is not to exhaust the subject; our idea is that we can contribute to theoretical enrichment and help with future research. 2018 marks exactly 140 years since the first influx of migrants to this region. We are still in our infancy.

REFERENCES

ACRE, State Government. Acre State Technology Foundation. **Atlas of the State of Acre / Acre State Government.** Acre State Technology Foundation. Rio Branco: FUNTAC, 2008.

. **Ecological-Economic Zoning of the State of Acre.** Phase II (Scale 1:250,000) - Synthesis Document. 2 ed. Rio Branco: SEMA, 2010. 356p.

. **ACRE IN NUMBERS 2009:** Rio Branco: SEPLAN, 2009.

ALBERTI, Verena. Stories within history. In: PINSKY, Carla Bassanezi (Org.). **Fontes históricas.** São Paulo: Contexto, 2005.

ALMA ACREANA. **Philosophy - Literature - Acreanities.** 2012. Available at: https://almaacreana.bloqspot.com.br/2016/02/53-fotos-raras-do-acre-antigo.html. Accessed in July 2017.

BACHELARD, Gaston. **Epistemology.** Rio de Janeiro: Zahar, 1983.

BAKHTIN, Mikhail. **Marxism and philosophy of language.** São Paulo: Hucitec, 1999.

BARBOSA, J. A. A psychology of the oppressed. In E. Bosi, **MEMÓRIA E SOCIEDADE:** lembranças de velhos (3rd ed., pp. 11-15). São Paulo: Companhia das Letras, 1994. p.12

BECKER, Olga Maria Schild. **SPACE MOBILITY OF THE POPULATION:** Concepts, Typology, Contexts. Rio de Janeiro: Art Line. 1997.

BELLO, Angela Ales. Intrapersonal and interpersonal. General lines of a philosophical-phenomenological anthropology. In: (org.) Juvenal Savian Filho. **EMPATHY EDMUND HUSSERL AND EDITH STEIN:** Didactic presentations. São Paulo: Edições Loyola, 2014. p. 09-28.

BENCHIMOL, Samuel. **AMAZONIA:** A Little Before and Beyond. Manaus: Umberto Calderaro, 1977.

. **Amazônia, um pouco-antes e além-depois.** Ufam 2ª . Ed. Manaus: revised, 2010

. **AMAZON:** War in the Forest. Manaus: EDUA, 2011.214p.

BHABHA, Homi K. **O local da cultura.** Belo Horizonte: Ed. UFMG, 1998.

BOGDAN, R., Biklen, S. **Qualitative Research in Education - an introduction to theory and methods.** Porto: Porto Editora, 1994.

BOLLNOW, Otto Friedrich. **Man and space.** Curitiba: UFPR. 2008.

BOSI, Ecléa. **Memória e Sociedade; Lembrança de velhos.** São Paulo: T.A. Queiroz, 1983.

BOURDIEU, Pierre. **Symbolic Power.** Lisbon: DIFEL, Difusão Editorial, 1989.

BRAZIL. IBGE - Brazilian Institute of Geography and Statistics. **2010 Demographic Census.** Available at: https://cidades.ibqe.qov.br/brasil/ac. Accessed in May 2018.

. **Brazilian Institute of Geography and Statistics.** 2018. Available at: https://cidades.ibqe.qov.br/brasil/ac/panorama. Accessed May 2018.

CALIXTO, Valdir. **Plácido de Castro and the Construction of Order in Aquiri:** contributions to the history of political ideas. Rio Branco: FEM, 2003.

CASSIANO, Luiz de Carvalho. **MARCH TO THE WEST:** an itinerary for the Estado Novo (1937-1945). 2002. Master's dissertation in History, Federal University of Brasília. Brasília: UnB, 2002.

CASTELO BRANCO, José M. B. **Povoamento da Acreania.** IHGB Magazine, volume 250. Rio de Janeiro, 1961.

DAMIANI, A. Luiza. The geography of population in "classical" geography. In. **Population and Geography.** São Paulo: Contexto, 1991.

DARDEL, E. L. **MAN AND LAND: The** Nature of Geographical Reality. Translation: Werther Holzer.

São Paulo: Perspectiva, 2011.

.   **L'Homme et la terre: nature de la realite geographique.** Paris: Presses Universitaires de France, 1952. 133p.

GAUDEMAR, Jean Paul de. **Mobility of Labour and Accumulation of Capital.** Lisbon: Estampa, 1977.

GIL, Antônio Carlos. **Methods and Techniques of Social Research.** 4 ed.; São Paulo: Atlas, 1994

.   **Methods and techniques of** social **research** / 6. ed. - 5. reimp. - São Paulo: Atlas, 2012.

GOETTERT J. D. & MORAIS M. J. **GEOGRAPHIES OF PASSAGE:** considerations on the before, during and after migration from the Northeast to the Amazon based on the writings of Samuel Benchimol. Niterói: Anais Jalla, 2010.

GOMES, P. C. da Costa. **Geography and modernity.** Rio de Janeiro: Bertrand Brasil, 1996.

GONÇALVES, R. C.; LISBOA, T. R. On the method of oral history in its life trajectories modality. **Rev. Katál.** Florianopolis v.10 n. esp. P. 83-92, 2007. Available at: file:///C:/Users/Master%20Liziane/Downloads/1145-19097-1-PB.pdf. Accessed in September 2016.

GUERREIRO, Luís. **Rubber in anti-seismic design.** FEUP, December 2003.   Available at :

http://www.globalconstroi.com/images/stories/Artigos_Tecnicos/Novos_materiaislll/B_orracha.pdf. Accessed February 2018.

HALBWACHES, Maurice. **Collective memory. "Collective memory and individual memory".** São Paulo: Vértice/Revista dos Tribunais, 1990.

HUSSERL, Edmund. **The idea of phenomenology.** Rio de Janeiro: Edições 70, 1989.

JAPIASSU, H.; MARCONDES, D. **Dicionário básico de Filosofia.** 3ª ed. Rio de Janeiro: Zahar, 2001.

LEFEBVRE, Henri. **The production of space** / Translated by Doralice Barros Pereira and Sérgio Martins (from the original: *La production de 1'espace.* 4e éd. Paris: Éditions Anthropos, 2000). First version: beginning - Feb 2006

LEFÈVRE, F. & LEFÈVRE, A. M. **The collective subject that speaks. Interface -** Comunicação, Saúde, Educação, 10(20), 517-524, 2006.

LEFÈVRE, F. **Discurso do Sujeito Coletivo:** um novo enfoque em pesquisa qualitativa. Caxias do Sul - RS: EDUCS, 2003.

LIMA, C. de Araújo. **PLÁCIDO DE CASTRO:** A caudillo against imperialism. Rio de Janeiro:

Brazilian Civilisation, 1973.

LOWENTHAL, David. **How We Know the Past.** History Project (17). São Paulo: EDUC, 1981.

MAIA, N. D. Silva. **The formative trajectory of Jean Pierre Chabloz. Art, rationality and intuition.** New Academic Editions: NEA, 2014.

MARTINELLO, Pedro. **The "Battle of the Rubber" in the Second World War.** Rio Branco: EDUFAC, 2004.

MEDEIROS FILHO, João; SOUZA, Itamar. **Os Degredados Filhos da Seca.** Petrópolis: Vozes, 1984.

MELO, Fabio. **Cipriano Barata Americanist Study Group.** 2013. Available at: http://geaciprianobarata.blogspot.com.br/2015/08/a-revolucao-acreana-1899-1903.html. Accessed July 2017

MELLO, Mauro, P. de. **The Boundary Question between the States of Acre, Amazonas and Rondônia.** Brazilian Journal of Geography. V52, n. 04. Rio de Janeiro, 1990.

MENEZES, F. L. Migration: a psychological perspective, a postmodern reading or simply a prejudiced view. In: CUNHA, M. J. C.; GURAN, M.; HASSE,

G.; MENEZES, F. L.; STEVENS, C. M. T.; **Migration and Identity:** perspectives on the theme. São Paulo: Centauro, 2007.

MERLEAU-PONTY, Maurice. **Phenomenology of perception.** Translated by: Carlos Alberto Ribeiro de Moura. 2ª . ed. São Paulo: Martins Fontes, 1999.

MONDARDO, Marcos Leandro. **THE PERIODS OF MIGRATION:** Territories and Identities in Francisco Beltrão/PR. Dourados-MS, 2009.

MORAES A. C. Albuquerque de. **JEAN-PIERRE CHABLOZ E A CAMPANHA DE MOBILIZAÇÃO DE TRABALHADORES PARA A AMAZÔNIA (1943):** poster and preliminary study in confrontation. VI EHA - Art History Meeting - UNICAMP 2010. Available at : http://www.unicamp.br/chaa/eha/atas/2010/ana_carolina_albuquerque.pdf. Accessed February 2018.

MORAES, M. J. THE CONSTRUCTION OF "URBAN CIVISM" IN THE CITY OF RIO BRANCO-AC: the discourse of "acreanity" and the territorial patrimonialisation of the capital of the State of Acre. In:. National Meeting of Geographers. **Electronic Proceedings.** N. 16°, Porto Alegre, 2010. Available at: http://www.aqb.orq.br/evento/download.php?idTrabalho=2125. Accessed July 2017.

. **"Acreanity": Invention and Reinvention of Acrean Identity.** 2008. PhD Thesis in Geography.

Fluminense Federal University. Niterói-RJ, December 2008.

MOREIRA, Ruy. The Technical Periods and the Paradigms of Labour Space. In:. **Geographical Science Magazine.** Year VI, N. 16, Vol. II, May - August. Bauru: AGB, 2000.

MORIN, E. **Introduction to complex thinking.** Porto Alegre: Sulina, 2005.

NASCIMENTO SILVA. M. G. S. N. **O espaço ribeirinho.** São Paulo: Terceira Margem, 2000.

OLIVEIRA, Luiz Antonio Pinto de. **O SERTANEJO, O BRABO E O POSSEIRO:** Os cem anos de wandanças da população acreana. Rio Branco: Acre State Government, 1985.

PELIANO, José Carlos. **Labour Accumulation and Capital Mobility.** Brasília: UNB, 1990.

PEREIRA, L. A. G.; CORREIA, I. S. PHENOMENOLOGICAL GEOGRAPHY: Space and Perception. **Caminhos de Geografia magazine.** September 2010. Available at: http://www.iq.ufu.br/revista/caminhos.html. Accessed December 2016

PINTO, Pimentel Júlio. **A MEMORY OF THE WORLD:** fiction, memory and history in Jorge Luís Borges. São Paulo: Estação Liberdade, 1998. 307p.

POULET, G. **The Proustian Space.** Rio de Janeiro: Imago, 1992.

PÓVOA NETO, Hélion. **Internal migration and labour mobility in present-day Brazil. New challenges for analysis.** Experimental, n.02, March 1997.

QUEIROZ, Maria I. P. de. **Variação sobre a Técnica** de **gravador no registro da informação viva. -** São Paulo: T. A. Biblioteca básica de ciências sociais. Series 2. Texts; v.7, 1991.

REIS, Arthur C. F. **O Seringai e o Seringueiro.** Rio de Janeiro: Ministry of Agriculture, 1953.

SALES, F. S.; SOUZA, F.C. de; JOHN, V. M. **The use of the DSC (Discourse of the Collective Subject) approach in educational research.** LINHAS, Florianópolis, v. 8, n. 1, jan. / jun. 2007.

SALIM, Celso A. "MIGRATION: The Fact and the Theoretical Controversy". In:. VIII National Meeting of Population Studies. **Electronic Proceedings,** vol. 3, São Paulo, ABEP, 1992.

SANTANA, Marquelino. SOLDIERS OF RUBBER - DIP propaganda and the Swiss painter who helped persuade northeasterners to come to the Amazon. **Blog Rondoniaovivo.**   2012. available at   : http://www.rondoniaovivo.com/noticias/soldados-da-borracha-a-propaganda-do-dip-   and-the-swedish-painter-who-helped-persuade-northeasterners-to-come-to-the-amazon/90534.   Accessed in July 2017.

SANTOS, A. C. de A. **ORAAL SOURCES:** Testimonies, life trajectories and history. Curitiba: DAP, 2005 p.3

SILVA, J. R. da. **Rubber Plantation Networks and the Spatial Organisation of Fortaleza do Abunã/Amazonia.** 2010.190p. Master's dissertation in Geography: Federal University of Rondônia Foundation - UNIR. Porto Velho, RO, 2010.

SINGER, Paul. **Political Economy of Urbanisation.** São Paulo: Contexto, 1998.

SILVA, Ronaldo da. **BRASILIA AND WASHINGTON:** Divergent Foreign Policy and the Prospects for South American Integration. 2010. PhD Thesis in Geography. Federal University of Uberlândia. Uberlândia MG, 2010.

SOUZA, Itamar de. **Internal Migrations in Brazil.** Rio de Janeiro: Vozes, 1980

SPOSITO, Eliseu Savério. **GEOGRAPHY AND PHILOSOPHY: A** contribution to the teaching of geographical thinking. 3ª reprint, São Paulo/SP: UNESP, 2004.

THOMPSON, P. **The voice of the past.** São Paulo: Paz e Terra, 1992.

TOCANTINS, Leandro. **Historical Formation of Acre.** Brasília: Senado Federal, V.1, 2001a.

TUAN, Yi-Fu. **SPACE AND PLACE:** The perspective of experience. Translation by Lívia de Oliveira. - São Paulo: DIFEL, 1983.

VIDAL, A. R. de N.; ALVES, F. C. D. Analysis of Regional Accounts 2010-2014. **ETENE Sectoral Notebook.** Year 0, n.4, April 2017. Available at: https://www.bnb.gov.br/documents/80223Z1722440/contas+regionais.pdf/e07a79c6- dbca-96ad-05b1-9f28b2501a52.accessed in July 2017.

VINUTO, J. **Snowball sampling in qualitative research: an open debate.** Temáticas, Campinas, v.22, n.44, p.203-220, Aug/Dec. 2014.

ZILLES, Urbano. Husserlian phenomenology as a radical method. In:. HUSSERL, Edmund. **The crisis of European humanity and philosophy.** 3 ed. Porto Alegre: EDIPUCRS, 2008.

Sites visited:

ACRE. **History and Geography of the State of Acre.** 2017. Available at: http://minas-cidades.blogspot.com.br/2014/09/acre-historia-e-geografia-do-estado-do.html. Accessed May 2018

MACHADO, Altino. History of Acre. **Altino Machado's blog.** 2005. Available at: http://www.altinomachado.com.br/2005/12/. Accessed July 2017

WE AND GEOGRAPHY, http://noss2geografia.blogspot.com.br/2012/02/o-territorio- legalised-the-treated.html. Accessed in July 2017.

PONTES, C. J. de F. "EL-DOURADO VERDE": The Acre War. **South American Journal of Basic**

Education, Technical and Technological. Vol.3, n. 1, 2015. Available at: file:///C:/Users/Master%20Liziane/Downloads/384-1440-1 - PB%20(1).pdf

PONTES, C. J. de F. O **PRIMEIRO CICLO DA BORRACHA NO ACRE:** Da Formação dos Seringais Ao Grande Colapso. Colégio de Aplicação da Universidade Federal do Acre. Vol.1, p.107-123, [Rio Branco], 2014. Available at: file:///C:/Users/Master%20Liziane/Downloads/100-279-1-PB.pdf. Accessed in July 2017.

APPENDIX

INTERVIEW SCRIPT

FIELDWORK 2017

1) Personal data (identification)

1.1 Name:

1.2 Age:

1.3 Residence:

2) Life trajectory (origins and migration) through the main issues:

2.1 Why did your family come to Acre?

2.2 What was the situation like in the rubber plantations?

2.3 What was the food like in the rubber plantations?

2.4 What are the dangers of attacks by wild animals?

2.5 Were there many diseases in the rubber plantations? What did you do in case of illness?

2.6 How did your wife and children help you in the rubber plantations?

2.7 Did you manage to study?

2.8 Did you manage to have any contact with your family in the north-east after you came to Acre?

2.9 Did they make a lot of money from cutting the syringe?

2.10 Compare before with now. Do you think it's improved? Which of these times do you prefer?

ANNEXES:[7]

---

[7] The photos and personal details of the interviewees published here are only of those from whom I obtained the life narratives and who authorised me to publish them in this work.

Former rubber tappers living in the state of Acre who took part in this research

Photo 03: Elvaldo Lima and his mother Rosa Lima, residents of Mâncio Lima, Acre.
Photo 04: On the left, Maria Liziane S. Silva. Right: Murilo de Lima, 98, from Ceará, who arrived in 1944. Resident of Mâncio Lima, Acre. Died in October 2018.
Photo 05: Mr "Lú" Mourão (as he is known), 79, resident of Mâncio Lima, Acre.

Photo: 06: Divaldo Alves de Souza, 92, and his wife Oscarina Alves de Souza, 90. Former rubber tappers from the municipality of Tarauacá, they now live in Rio Branco. Acre.

Photo 07: Maria Helena, 87, and her husband Manoel Claudio, 81. Residents of Mâncio Lima, Acre.
Photo 08: Dona "Inha" (as she is known) and her husband Adalton M. de Lima, 83. Residents of Mâncio Lima, Acre.
Photo 09: Dona "Dê" (as she was known). She died five months after the interview, aged 84. She lived in Mâncio, Acre.

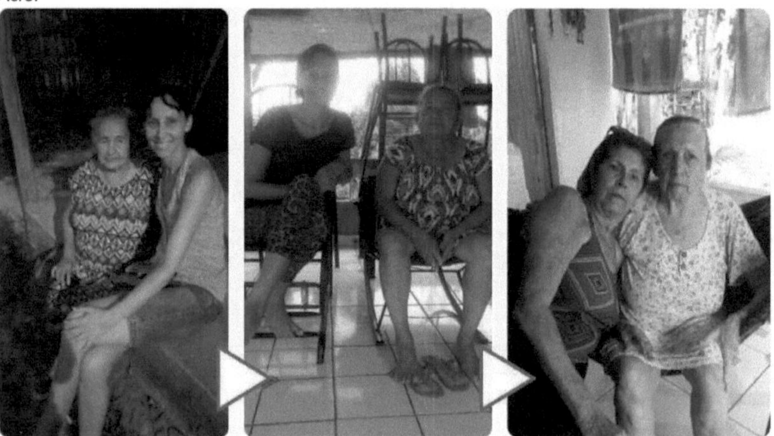

Photo: 10: Maria dos Anjos, 86. Resident of Mâncio Lima, Acre.
Photo 11: Osmarina, 72, resident of Mâncio Lima, Acre.
Photo: 12: Right: Zilmar Marques, daughter of 91-year-old Guiomar Medeiro Marques. Residents of Mâncio Lima, Acre.

Printed by Books on Demand GmbH, Norderstedt / Germany